Fault Tolerant Drive By Wire Systems: Impact on Vehicle Safety and Reliability

Edited by

Sohel Anwar

Indiana University Purdue University Indianapolis
USA

eBooks End User License Agreement

CONTENTS

CHAPTERS

FOREWORD

High-integrity systems require a comprehensive fault tolerance for the components and the corresponding control system. This includes the design of fault-tolerant sensors, actuators, process parts, computers, communication (bus) systems and control-algorithms. The fault tolerance is frequently based on redundancy in hardware and also software, either as static redundancy with several modules or dynamic redundancy with cold or hot standby modules. Examples of such redundant systems are known for train and nuclear power systems, and for space vehicles and aircraft. Especially the fly-by-wire systems for aircraft demonstrate a very high state-of the art of hardware and software redundancy and fault tolerance.

Drive-by-wire systems for automobiles at the actuator level are since more than 15 years in series production for e.g. the electrical throttle, automatic transmissions, electro-hydraulic brakes and parking brakes. For steering the hydraulic power steering has been replaced by the electrical power steering up to middle class cars, but is still based on the mechanical steering column. This means, that for this very safety critical chassis component one relies on a mechanical back-up in case of a failure of the power support.

Compared to fly-by wire systems drive-by-wire systems cannot have the high level of redundancies because of costs, maintenance issues and the large production numbers with are use by non trained drivers, as compared to professional pilots, and any kind of environment and usage. However, in general context of the electrification of automobiles and advanced driver assistance systems the future will show how far drive-by-wire systems will develop. The modern driver assistance and chassis control systems like ABS, ESC, ACC and suspension control are already "by-wire" or "bycontrol software" on higher levels.

Therefore this eBook is a timely compilation of the present and future development of fault-tolerant drive-by-wire systems by having a focus on the important issues of safety and reliability. I wish Professor Sohel Anwar and the authors of this eBook a very good success.

Rolf Isermann

Darmstadt, January 2011

PREFACE

Drive-By-Wire systems are still a long way off from their full potential, primarily due to reliability and safety concerns. While throttle-by-wire systems are commonplace in many high-end vehicles, brake-by-wire and steer-by-wire systems are still in the research and development stage with major automakers. Daimler-Benz introduced the first brake-by-wire system using electronically controlled hydraulic valves and pumps in the SL 500 and E-Class models as early as 2001. However, the high cost of these brake systems and some reliability/safety issues did not permit widespread adoption of these systems by the automotive industry. Thus the major automakers, in general, were discouraged from investing further in the development of drive-by-wire systems. However, researchers at universities and government labs continued to search for better solution to improve the reliability and safety of drive-by-wire systems in a cost effective manner.

Overall safety and reliability of drive by wire systems can be significantly improved *via* redundancy based system architecture with fault tolerant control methodology. This type of hardware-software solutions have already been successfully developed and implemented in fly-by-wire systems. For drive-by-wire systems, however, the challenge is to design a hardware-software architecture for enhanced safety and reliability in a cost effective manner. This e-book aims at addressing some of these challenges via a number of approaches, such as, analytical redundancy, system level reliability enhancements, and global positioning system assisted steer by wire system.

Fault tolerant control of drive by wire systems is the focus of this eBook. The authors of this eBook are experts in the field of fault tolerant drive by wire systems. Chapter 1 of this eBook gives an overview of the drive by wire systems and a brief background on the challenges facing such system for commercial viability. Chapter 2 presents system level reliability and enhancements to drive by wire systems. Chapter 3 introduces the dependability and functional safety of drive by wire systems. Chapter 4 describes a GPS (Global Positioning System) aided steer-by-wire control system for articulated vehicles. Chapter 5 presents a virtual operator model for construction equipment design. All of these chapters capture various aspects of safety and reliability of drive by wire systems which is the most challenging question today for such systems.

This eBook can be used as a reference book or as a textbook for a graduate course in the area of Safety and Reliability in the context of automotive control systems. It also offers some possibilities of further developments including important problems in this research area.

Sohel Anwar

Indiana University Purdue University Indianapoli
USA

List of Contributors

Sohel Anwar Department of Mechanical Engineering, Purdue School of Engineering and Technology, Indiana University Purdue University Indianapolis, 723 W. Michigan Street, SL 260N, Indianapolis, Indiana 46202, USA

E-mail: soanwar@iupui.edu

Giuseppe Buja Department of Electrical Engineering, University of Padova, *Via* Gradenigo 6/a, 35131 Padova, Italy

E-mail: giuseppe.buja@unipd.it

Sabri Cetinkunt Department of Mechanical and Industrial Engineering, University of Illinois at Chicago, 842 W. Taylor Street (MC 251), Chicago, IL 60607, USA

E-mail: scetin@uic.edu

Ahmed Adel Elezaby University of Illinois at Chicago, 842 W. Taylor Street (MC 251), Chicago, IL 60607, USA

E-mail: aeleza2@uic.edu

M. Abul Masrur US Army RDECOM-TARDEC, RDTA-RS, MS-233, 6501 E. 11 Mile Road, Warren, MI 48397-5000, USA

E-mail: md.abul.masrur@us.army.mil

Roberto Menis Department of Electrotechnics, Electronics and Computer Science, University of Trieste, *Via* Valerio 10, 34127 Trieste, Italy

E-mail: menis@units.it

Rami Nasrallah Caterpillar, Inc., Peoria, IL, USA

CHAPTER 1

Introduction and Overview of Fault Tolerant Drive by Wire Systems

Sohel Anwar[*]

Indiana University Purdue University Indianapolis, USA

Abstract: A brief overview of fault tolerant drive-by-wire technology is presented in this chapter. A review of drive by wire system benefits in performance enhancements and vehicle active safety is then discussed. Challenges in relation to the fault tolerant design of drive by wire system are then presented. This is followed by in-depth coverage of fault tolerant design of a steer by wire system. Future trends in the design of fault tolerant drive by wire systems are presented at the conclusion of the chapter.

Keywords: Drive By Wire, Brake By Wire, Steer By Wire, Throttle By Wire, Fault Tolerant Control, Sliding Mode Observer, Generalized Prediction, Static Redundancy, Dynamic Redundancy, Analytical Redundancy, Fault Tolerant Communication.

INTRODUCTION

A drive by wire (DBW) system is an automotive system that interprets driver's inputs and executes the commands to produce desired vehicle behavior, typically *via* a microprocessor-based control system. A typical drive-by-wire system comprises of redundant sensors, actuators, microprocessors, and communication channels for fault tolerance. There are no mechanical or hydraulic connections between driver's input interface (e.g. throttle, brake, steering) and vehicle system (e.g. engine/traction motor, brake/steering actuators) in a drive by wire equipped vehicle.

The first true drive by wire system to come to the market was Throttle By Wire (TBW) which was incorporated in high end vehicles such as Audi A6, Mercedes Benz, Lexus, and BMW models in the late 1990's and early 2000's. The TBW systems were advantageous in stability control applications where the throttle deactivation may be needed in order to improve the traction so that sufficient brake torque can be generated.

Electro-hydraulic brake (EHB) system, a form of brake by wire (BBW), was first introduced in Mercedes Benz SL series in 2001-02 (Higgins & Koucky, 2002). Although hydraulically actuated, these brakes operate on commands from sensors at the brake pedal and generate the necessary brake pressure at the wheel cylinders *via* a set of electronically controlled valves and a pump. However, the brake by wire system was decommissioned and removed from the vehicle due to a number of field problems a few years later. Work on the electro-mechanical brakes (EMB), another form of brake by wire system that does not use hydraulic fluid, was done in the late 1990's by number of automotive companies such as Bosch, Continental, and TRW. However, issues related to their reliability and fault tolerance still remain which must be addressed before these system can be used in an automobile.

Steer by wire (SBW) system is by far the most complex drive by wire system which is also the most safety critical by-wire system in an automobile. In a pure steer by wire system, the steering column is eliminated. Sensors mounted on the steering wheel are interpreted by the controller to generate the correct amount of road wheel angle using electric motors based on the vehicle velocity. If a sensor stops functioning properly, the controller will not be able to actuate the motors to generate the correct road wheel angle, potentially causing hazardous situation.

---

*Address correspondence to Sohel Anwar: Department of Mechanical Engineering, Purdue School of Engineering and Technology, Indiana University Purdue University Indianapolis, 723 W. Michigan Street, SL 260N, Indianapolis, Indiana 46202, USA; E-mail: soanwar@iupui.edu

In a broader definition, hybrid electric vehicles, electric vehicles, and plug-in hybrid electric vehicles can also be classified as drive by wire equipped automobiles due to the electronic control of various subsystems in these vehicles. Electric vehicles (EV) by their very nature are drive by wire that is propelled by electronic control of the electric traction motor based on the sensor information from the throttle pedal. However, the steering and brakes of an EV may still be hydro-mechanically operated. In case of hybrid electric vehicle (HEV), a sophisticated microprocessor based control system channels the power flow between the internal combustion (IC) engine, the battery, the electric motor/generator, and the vehicle wheels (Lu & Hedrick, 2005). All of these functions are done *via* a central controller for optimal performance. Plug-in hybrid electric vehicles are very similar to hybrid electric vehicle, except that a more powerful battery extends the vehicle range in pure electric mode.

This chapter is organized as follows: A more detailed coverage on fault tolerant drive by wire system is covered in the next section. The performance and safety benefits of the drive by wire systems are illustrated in the following section. This is followed by the section on technological challenges and possible solutions associated with DBW system. Future trends for the DBW system is presented in the last section.

DRIVE BY WIRE SYSTEMS – FAULT TOLERANT DESIGN

Fig. (**1**) shows a pictorial view of a number of drive by wire systems in a concept automobile. With all the drive by wire systems in the vehicle which use electrical power for actuation, the electric power demand must be met using higher voltage systems The power demand for hybrid electric vehicles is significantly higher due to propulsion need and hence uses a 240-300V DC power bus. An extra layer of safety in the design of such systems must be incorporated to eliminate the possibility of electrocution. In addition, for fault tolerant architecture, at least two or more such power sources are required for a DBW equipped vehicle.

The first commercialized brake by wire system was introduced by Daimler Benz (Higgins & Koucky, 2002). This is an electro-hydraulic brake (EHB) system with mechanical backup that was installed in SL 500 model. The brake pedal displacement sensor output is used to determine the desired wheel cylinder pressure which is generated *via* a closed loop control system that includes a set of electro-hydraulic valves and an electric motor driven pump. This system was later recalled due to reliability issues. Since then no automakers have incorporated brake by wire systems in any of their vehicles.

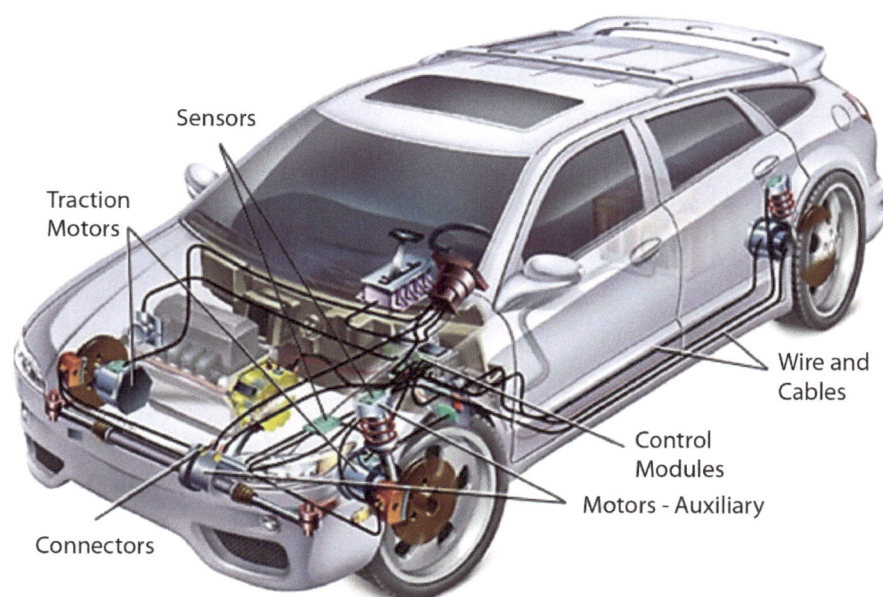

Fig. (1). A drive by wire equipped hybrid electric concept vehicle.

This concept vehicle in Fig. (**1**) is based on the synergy of combining a DBW system with HEV. The DBW systems can easily be powered by the HEV battery pack or the high voltage power bus. The IC engine along with the motor generator will ensure that power is always available for the DBW systems.

Drive-By-Wire systems offer a number of benefits when incorporated on a vehicle. Some of the benefits are as follows:

- DBW systems can easily be configured (*via* software updates) for added or tunable features such as brake pedal feel/enhanced safety *via* stability control.

- DBW systems can enhance vehicle performance by prepositioning brake calipers for fast brake actuation or allowing for oversteering to enhance maneuverability.

- DBW systems can improve fuel economy *via* better engine/motor/powertrain control and *via* regenerative braking.

- DBW systems can offer better ergonomics such as adjustable feel at driver's interface (steering wheel, brake/accelerator pedals).

- DBW systems offer better fault detection and warning to the driver which enhances the safety, reliability, and maintenance of the vehicle.

- DBW systems are able to incorporate multi-functionality in a single system thereby making today's advanced features (e.g. stability control systems *via* active steering) more cost effective in these automobiles.

- SBW systems can not only provide these additional features, but also can free up premium packaging space by eliminating the steering column thereby enabling easy assembly of the instrument panel.

- SBW systems also offer variable steering ratio at different vehicle speeds for additional safety.

- Automatic line keeping is possible for SBW equipped vehicle on the automated highways of the future. This feature will further enhance safety and comfort of the driver.

- Haptic steering for improved ergonomics and operator safety & performance is possible with SBW vehicles.

- DBW systems will offer reduced overall production cost for steering systems *via* standardized modules and software.

- BBW systems offers improved brake noise, vibration, and harshness (NVH) since there is no direct mechanical or hydraulic link from the pedal to the wheels. ABS pulsation in normal hydraulic brake will disappear in a BBW system.

- BBW systems also allows for Better Comfort Level

  - Anti-Dive Algorithm

- Better fuel economy through regenerative braking

- Easier maintenance

  - Less wear and tear

  - No need to turn those rotors

- Extended brake life

  - No need to replace brake shoes, pads, or rotors

- Better cold performance than wet brake systems

- Better braking-in-a-turn performance

- Enhanced safety

  - Stability control algorithms

  - Reduction of brake fade

- Enhanced cruise control on a down hill

However, there are a number of challenges that a DBW system must address before full commercialization. These challenges are discussed in the "Technological Challenges" section.

THROTTLE BY WIRE (TBW) SYSTEMS

For traction control system based on engine intervention, it is necessary to control engine torque. By electronically controlling the mass air flow rate *via* a throttle by wire (TBW) system, it is possible to achieve this functionality. For yaw stability control system, it may be necessary to control the engine torque to the wheels to maximize available traction. TBW systems allow this functionality through their very design. For hybrid electric vehicle propulsion, it is necessary to mechanically disconnect the pedal from the throttle to allow for torque blending between the engine and electric motor which can easily be achieved *via* a TBW system. A TBW system also provides an easier way to implement engine braking for commercial vehicle (Jake brake, Exhaust gas brake, etc.). For autonomous vehicles, it is necessary to have a TBW system for vehicle propulsion.

The basic operation of a TBW system is that the pedal sensor information along with throttle angle feedback is used to actuate the TBW motor to maintain the commanded throttle angle. Fig. (2) shows an electronic throttle body (ETB) used in a TBW system. As indicated earlier, a TBW system is a natural fit in an active safety system or a combination of active systems (as in this case) where upon the ABS/TCS/YSC microprocessors share information with a TBW microprocessor and a supervisory microprocessor. The TBW system microprocessor gives preference to the active safety throttle command over its own pedal command and controls the throttle angle accordingly.

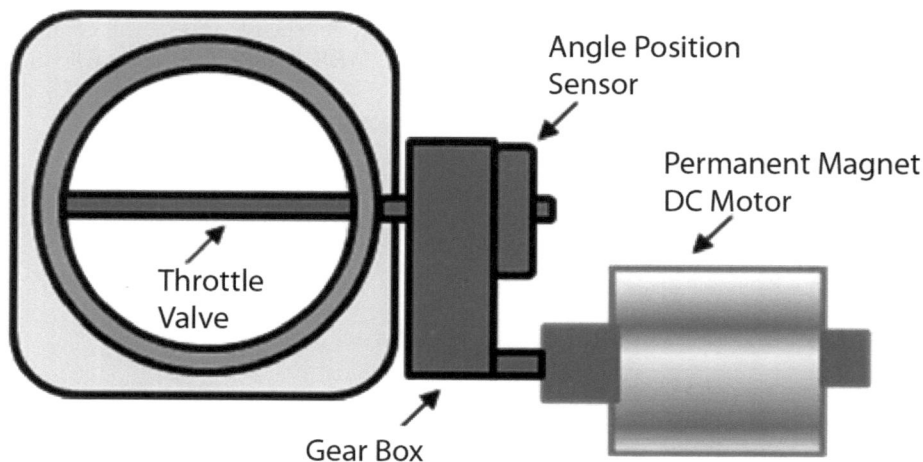

Fig. (2). Schematic of an electronic throttle body in a TBW system.

BRAKE BY WIRE (BBW) SYSTEMS

Brake by wire systems can be classified in three main categories, based on the type of actuation system, as stated below:

1.   Electro-hydraulic (EH) BBW systems

2.   Electromechanical (EM) BBW systems

3.   Electromagnetic (EMg) BBW systems (Regenerative braking *via* electric generator or Eddy Current braking)

The electro-hydraulic brake by wire system is composed of an electronically controlled pump/motor assembly, a set electronically controlled valves, pressure transducers, brake pedal sensors, driver modules, and microcontrollers. Fig. (**3a**) illustrates the actuation system of an EH brake by wire system. The brake pedal sensor information is sent to microcontroller which interprets this in terms of a desired wheel cylinder pressure command. The microcontroller also receives the wheel cylinder pressure information from the pressure transducer. Based on the desired and actual wheel cylinder pressure values, the microcontroller calculates appropriate pump and valve commands (*via* a control algorithm) which are then used to either build or dump the pressure in order to minimize the pressure error.

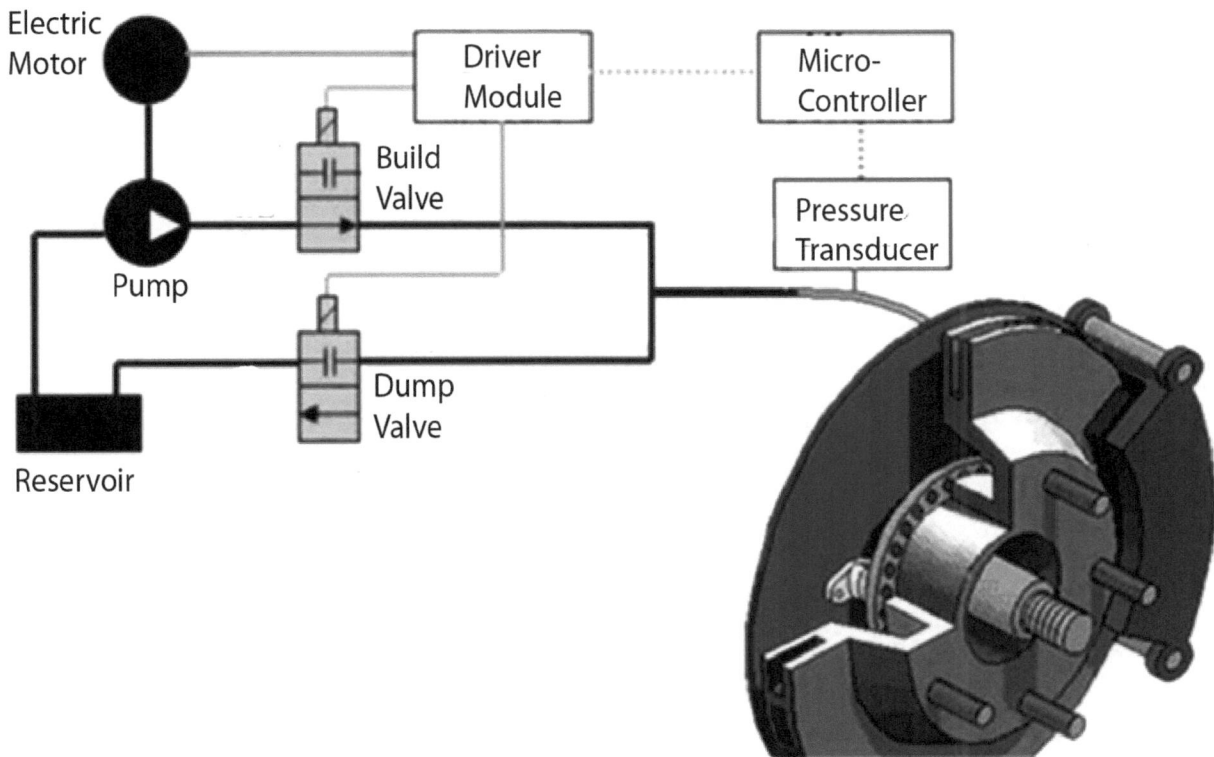

Fig. (3a). Schematic of an electro-hydraulic brake by wire actuation system.

The electromechanical (EM) brake by wire system is a dry system and no hydraulic fluid is used. It still uses the friction pads or shoes for braking the vehicle as in EH BBW system. However, the actuation mechanism is based on an electric motor and a mechanism that applies force onto the friction pads. Fig. (**3b**) illustrates the operation of this type of brake system (Hoseinnezhad, 2006). In addition to the brake pedal sensor, a number of other sensory information becomes necessary for such a system (e.g. clamp force, actuator position).

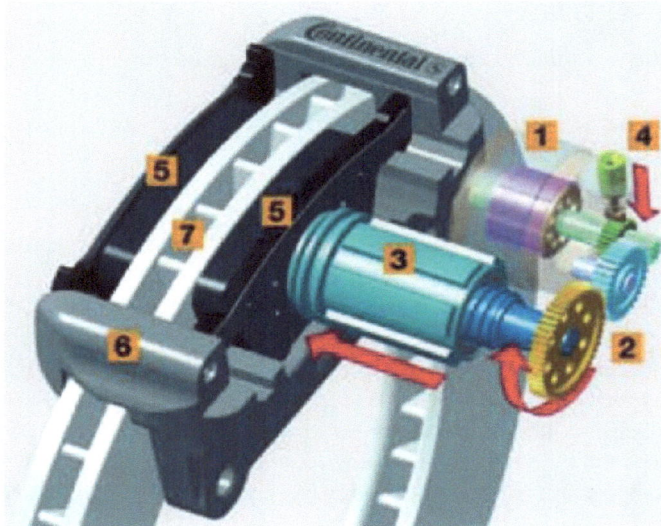

Fig. (3b). Electromechanical brake by wire actuation system.

The third type of brake by wire system is based on electromagnetic retardation. The principles of electromagnetic retardation are utilized to generate the braking forces which are contactless. As a result, no friction pad is needed in electromagnetic braking. Two principal types of actuators are used: i) Eddy current brakes, and 2) Regenerative braking *via* generator.

Fig. (3c) shows a picture of an eddy current braking actuator which is composed of a wire wound stator and solid iron rotor. When the stator is energized, an electromagnetic field is generated which gives rise to Lorentz forces according to Maxwell's equation when the rotor turns within the electromagnetic field. These forces oppose the motion of the rotor, thereby producing retarding torque. Regenerative braking works in a similar fashion. One limitation of electromagnetic braking systems is their dependency on the rotational speed, that is, the retarding torque is a function of the rotor speed. As a result, at lower speed, the retarding torque diminishes significantly and will not be able to provide sufficient braking torque. For this reason, electromagnetic brake by wire systems are rarely used as a standalone brake system. They are often used in conjunction with a friction brake system.

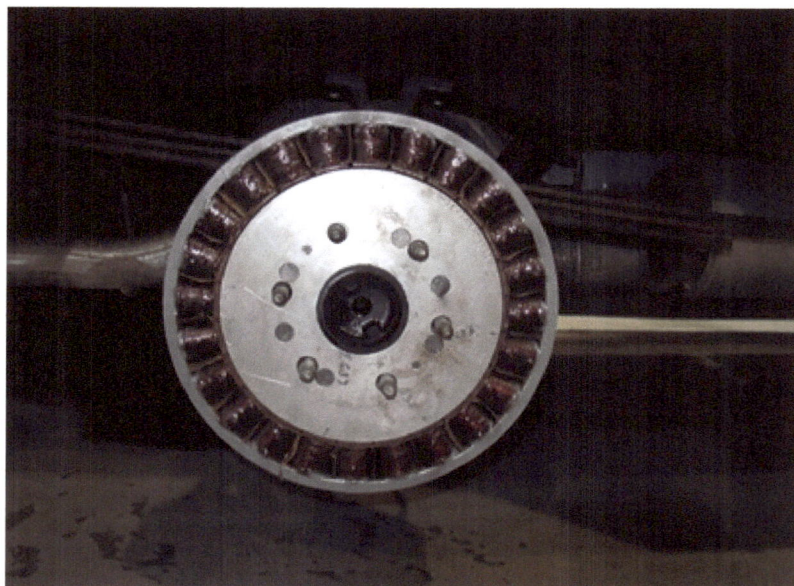

Fig. (3c). Electromagnetic brake by wire actuation system.

STEER BY WIRE (SBW) SYSTEMS

Steer by wire systems are in the early stage of development. Major automakers are yet to decide if and when such system will be incorporated in a vehicle. Similar to the brake by wire systems, SBW systems use microcontrollers to actuate the road wheel in the lateral direction based on the hand wheel steering angle and vehicle speed information. Fig. (**4**) illustrates a possible configuration of a steer by wire system. In this particular configuration, the hand wheel steering angle is measured *via* an appropriate sensor which is then communicated to two microcontrollers. These microcontrollers also receive vehicle speed information from the vehicle communication bus. Based on the steering wheel angle and vehicle speed information, the microcontrollers calculated the road wheel angle command *via* suitable algorithm. The microcontrollers then tries to maintain the commanded road wheel angle by actuating the rack mounted motors in a closed loop fashion *via* road wheel angle feedback.

Steer-By-Wire systems are a relatively new development compared to the traditional mechanical, hydraulic or electric steering systems that are currently used for motor vehicles. In the SBW system, there is no mechanical coupling between the steering wheel and the steering mechanism, i.e. the vehicle's steering wheel is detached from the steering mechanism during normal operation. Even though the mechanical linkage between the steering wheel and the road wheels has been eliminated, a SBW steering system is expected not only to implement the same functions as a conventional mechanically linked steering system, but also to provide the advanced steering functions.

There are a number of important steering functional requirements for a Steer-By-Wire system:

- Directional control and wheel synchronization.

- Adjustable variable steering feel.

- Adjustable steering wheel returns capability.

- Variable end of travel stop for steering wheel.

- Variable steering ratio.

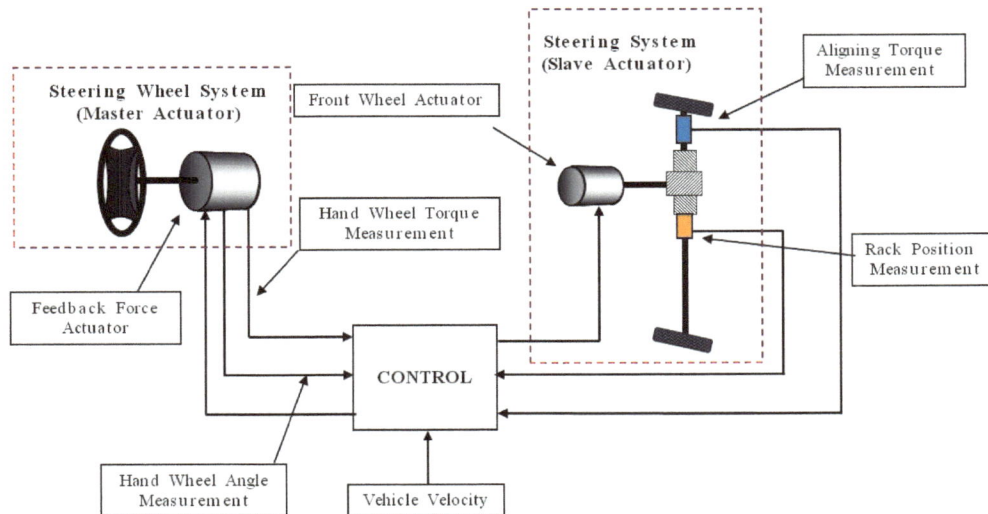

Fig. (4). Schematic of an SBW system.

It can be noticed that all of the drive by wire systems described above comprised of multiple redundancy in the sensors, actuator, microcontrollers, and communications buses. This is obviously by design which ensures a fault tolerant operation. Of the three drive by wire systems presented above, both BBW and SBW systems are considered safety critical systems which means that any failure of the system may lead to catastrophic consequences. As a result, these systems must be designed to withstand multiple faults or failures within the system, before reaching conditions of catastrophic proportions. A fault in sensor, actuator, or microcontroller must not cause any severe consequences. This fault tolerance feature requires multiple redundancy in the hardware architecture which provides fault tolerant operation.

PERFORMANCE ENHANCEMENTS AND ACTIVE SAFETY *VIA* DRIVE BY WIRE SYSTEMS

Drive by wire systems can significantly enhance the design and performance of the vehicle active safety systems such as the collision avoidance systems, adaptive cruise control, active front steering, stability control, road condition warning. Fig. (**5**) illustrates trend of incorporation of various active safety systems in automobiles and driver error mitigation *via* DBW systems. The more advanced systems (e.g. platooning, highway copilot, autonomous driving) can only be effectively achieved *via* drive by wire system. Other active safety systems are a natural fit for drive by wire systems as these systems can easily be incorporated *via* the addition of a software module in a drive by wire equipped vehicle. As described in (Anwar, 2006), the ABS algorithm was implemented *via* only software update to a hybrid brake by wire system. In addition to passenger vehicles, SBW systems have also been investigated in the context of large construction machineries such as wheel loaders with superior performance enhancements (Haggag *et al.*, 2005).

Drive by wire systems can also significantly enhance the performance of a vehicle by taking proactive actions in a driving scenario. For example, vehicle handling can be made more responsive by adding slight oversteer on high friction coefficient surface which can easily be done *via* a steer by wire system (Hebden *et al.*, 2004; Yih & Gerdes, 2005; Chang, 2007). Performance enhancement *via* SBW system while maintaining stability has also been investigated by a number of researchers (Limpibunterng & Fujioka, 2002; Oh *et al.*, 2004; Segawa *et al.*, 2004; Limpibunterng & Fujioka, 2004). Similarly vehicle suspension can be tuned dynamically *via* active suspension control system to adapt to road conditions the vehicle is in.

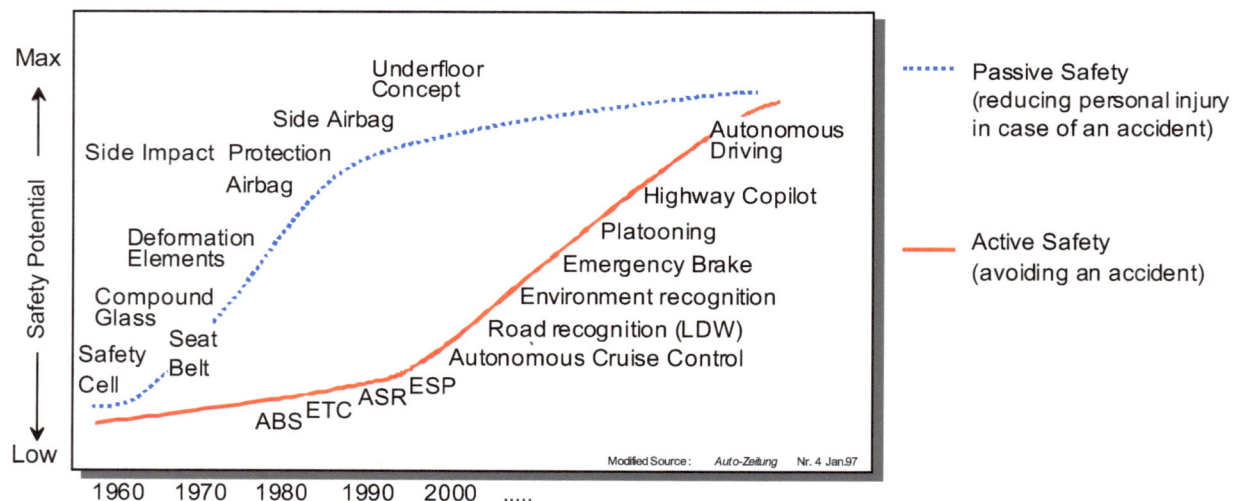

Fig. (5). Trend of active safety systems in automobiles.

Both brake-based Yaw Stability Control (YSC) and Anti-lock Brake System (ABS) have been demonstrated on BBW systems (Anwar, 2005; Anwar, 2006). Using sliding mode control (SMC) theory, the author provided experimental results for the hybrid BBW system. Here it was emphasized that the ABS system was implemented only through the software and no hardware modification was needed.

YSC algorithm *via* active front wheel steering has also been developed and tested in a SBW equipped vehicle (Yih & Gerdes, 2005; Zheng & Anwar, 2008). The test results from these studies indicate that the steering based YSC system on a SBW equipped vehicles has significant safety advantages in critical situations over a conventional vehicle where the driver has to manually control the vehicle against an unexpected yaw motion.

TECHNOLOGICAL CHALLENGES FOR DRIVE BY WIRE SYSTEMS

The complexity of the DBW systems arises from the fact that these systems must incorporate multiple redundant sensors, actuators, controllers, and communications networks to achieve fault tolerance. The fault tolerant control of these systems is accomplished *via* appropriate fault detection, isolation, and accommodation (FDIA) algorithms. However, the total number of redundant components makes the DBW systems prohibitively expensive. Also, in order to accurately detect and isolate a component failure without raising any false alarms, fast and robust detection algorithms are needed.

A number of issues must be resolved before drive by wire system equipped vehicles can be brought to the market. Cost and reliability are two major challenges facing these systems. While reliability of these systems can be improved *via* increased redundancy and fault tolerant control algorithms and communication protocols, the physical redundancy of components makes these systems prohibitively expensive. In this section, we highlight the major challenges facing the drive by wire systems including cost, reliability, packaging, control, and communication protocols.

The numerous benefits of DBW systems outweigh the risks in introducing such systems in the future automobiles. The concept vehicles with DBW systems have demonstrated such benefits (Stanton & Marsden, 1997). However, the overall cost of highly reliable DBW system is still several times higher than the conventional systems; primarily due to the presence of multitude of redundant components (sensors, microcontrollers, actuators etc.). Model-based fault detection techniques can be one possible solution, which could lower the overall cost without compromising reliability.

FAULT TOLERANT CONTROL

Fault tolerant control of a DBW system allows continual functionality of the DBW system in the event of failure of a component or subsystem. If operating quality decreases at all, the decrease is proportional to the severity of the failure, as compared to a naively-designed system in which even a small failure can cause total breakdown.

In order to bring down the cost, the total number of redundant components must be reduced without compromising the fault tolerance. One possible solution to this problem is to utilize analytical redundancy or model based fault detection, isolation, and accommodation. Model-based Fault Detection and Isolation (FDI) explicitly use a mathematical model of the system. It is motivated by the conviction that utilizing deeper knowledge of the system results in more reliable diagnostic decisions. The main idea is "analytical redundancy" which makes comparison of measurement data with known mathematical model of the physical process. It is superior to "hardware redundancy" generated by installing multiple sensors for the same measured variable. They offer simplicity, flexibility in the structure, less hardware, less weight, and cost.

Model-based Fault Detection and Isolation is achieved by implementing a more complex failure detection algorithm that takes careful account of system dynamics, which may be able to reduce requirements for costly hardware redundancy (Isermann, 2005; Isermann, 2006; Isermann & Munchhof, 2011). Analytical redundancy based FDI uses a model of the dynamic system to generate the redundancy required for failure detection. In many systems, all of the states cannot be measured because of cost, weight and size considerations, therefore, FDI schemes for such systems must extract the redundant information from dissimilar sensors, using the differential equations that relate their outputs. In addition to taking hardware issues into consideration, the designer needs to consider the issue of computational complexity. Most model-based FDI methods rely on analytical redundancy (Munchhof *et al.*, 2009a; Munchhof *et al.*, 2009b). In contrast to physical redundancy, when measurements from different sensors are compared, now sensory measurements are compared to analytically obtained values of the respective variable and the resulting differences are called residuals. The deviation of residuals from the ideal value of zero is the combined result of noise, modeling errors and faults. A logical pattern is generated showing which residuals can be considered normal and which ones indicate a fault. Such a pattern is called the signature of the fault. The final step of the procedure is the analysis of the logical patterns obtained from the residuals, with the aim of isolating the failures that cause them. Such analysis may be performed by comparison to a set of patterns known to belong to sample failures or by the use of some more complex logical procedure.

One major difference between pure DBW systems and other electronically controlled vehicle systems is that even if these control systems fail, the basic functionality of the brake and steering system remains intact. In case of DBW systems, a component failure can result in loss of functionality unless the DBW system has redundancy in its components by design. A DBW system with mechanical backup will allow a "limp home" mode in case of a component or subsystem failure. However, the mechanical backup is tantamount to having the basic mechanical system (e.g. steering or braking) and hence cannot be justified from a cost standpoint. Following the example of fly by wire systems, it is only natural to focus on drive by wire system without mechanical backup. However, DBW systems are microprocessor-based control system based on sensor inputs and electronically controlled actuation systems (*via* communication bus) which call for fault tolerance in case of component failure. Fig. (**6**) shows an example of the redundancy architecture of a fault tolerant drive by wire system (configuration similar to Isermann *et al.*, 2002).

Safety integrity is defined by the probability of a safety-related system satisfactorily performing the required safety functions under all the stated conditions within a stated period of time. Safety and reliability of a DBW system can generally be achieved by a combination of the following: fault avoidance, fault removal, fault tolerance, fault detection and diagnosis, automatic supervision and protection.

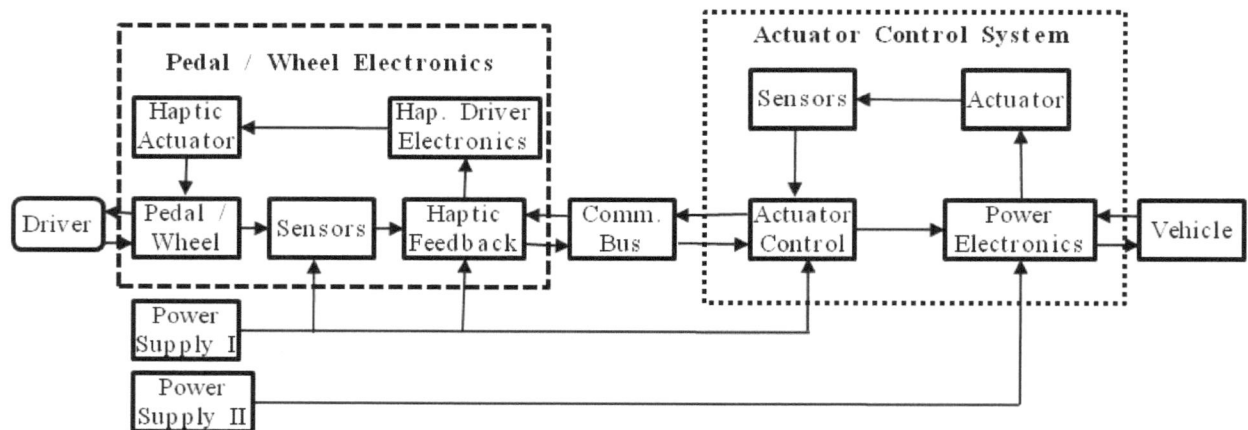

Fig. (6). Redundancy architecture of a drive by wire system.

Fault avoidance and removal has to be accomplished mainly during the design and testing phase. Analytical methods to evaluate the effects of faults on system reliability and safety exist: reliability analysis, fault tree analysis (FTA), failure modes and effects analysis (FMEA), hazard analysis (HA), and risk classification.

Certain component faults/failures can occur even after designing the hardware based on reliability and safety analysis. These faults must be tolerated, generally, by adding redundancy in the components, units, or subsystems. Redundancies in sensors, actuators, microprocessors, communication buses, power supplies are considered to achieve the fault tolerance. Two basic approached to fault tolerance: static redundancy and dynamic redundancy.

The fault in the main component is determined by consistency checking of its output signal (such as range of signal and its rate of change) and comparing with that from the standby component. For microprocessor, parity checking or watchdog timers may be used to detect the faulty microprocessor. After a fault is detected, a reconfiguration module then switches the standby component to take over and removes the faulty component from operation.

For static redundancy, at least three redundant component output signals are required for accurate fault detection *via* majority voting algorithm. In case of dynamic redundancy, model-based approaches can be utilized to detect fault since output signals are readily available from various components. Static redundancy can be found in various mechanical and electrical systems: spoke-wheel, multiple brushes in DC motor. Only single point hazardous failures are considered for mitigation upon fault detection. It has been shown that consideration of only single point failures is sufficient for a DBW system as opposed to a Fly By Wire (FBW) system.

Fault detection is an integral part of a fault tolerant control system. Fault detection methods based on measured signals can be classified as follows:

•   Limit value checking (thresholds) and plausibility checks ranges of single signals.

•   Signal-model-based methods for single periodic or stochastic signals.

•   Process-model-based methods for two or more related signals.

In order to obtain specific symptoms, it is necessary to have more than one input and one output signal for parity equations or output observers. For parameter estimation, it is necessary to have at least one input signal and one output signal. Fig. **(7)** illustrates the fault detection process that utilizes both signal model based fault detection and process model based fault detection (configuration similar to Isermann *et al.*, 2002).

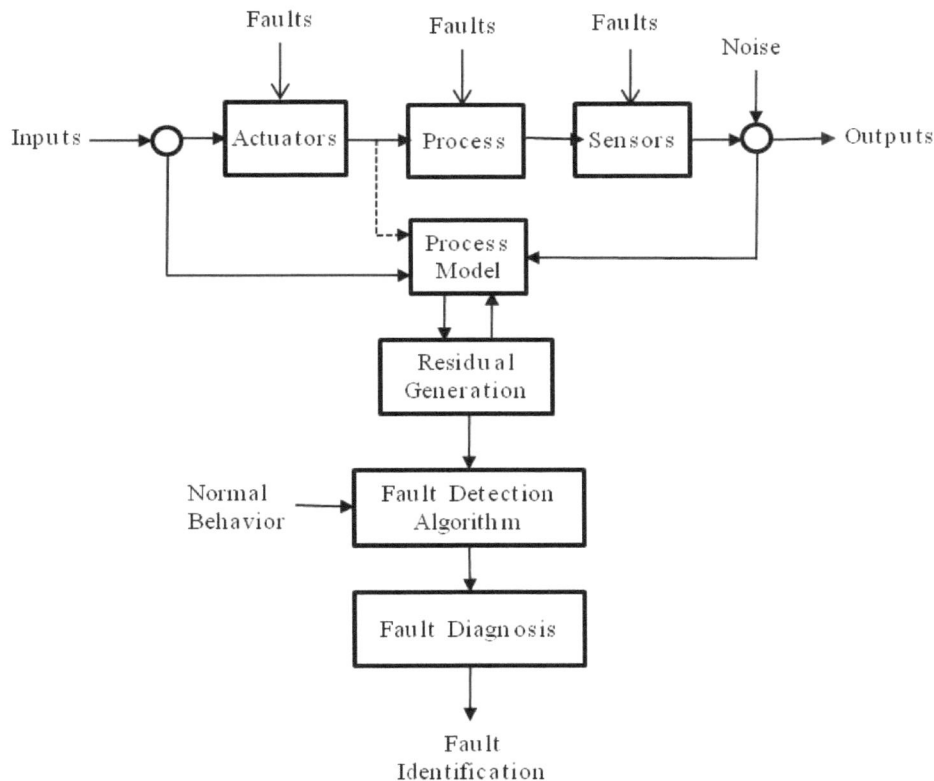

Fig. (7). Fault detection flow diagram.

Comprehensive overall fault-tolerance design can obtain fault tolerant components and fault tolerant control. This, in turn, can be achieved *via* fault tolerant lower cost components with built-in fault tolerance. A fault tolerant sensor configuration should be at least fail-operational after one fault. This can be obtained *via* hardware-redundancy with same type of sensor or by analytical redundancy with difference sensors and process models.

Fault tolerant actuators can be designed by using multiple actuators in parallel, with either static or dynamic redundancy with cold or hot standby, e.g. static redundancy in hydraulic actuators in a fly-by-wire system. Fault tolerant actuators can also be designed with redundancy in actuator components that have the lowest reliability, e.g. two servo valves for hydraulic actuators or three windings on an electric motor are examples of static redundancy. The cabin pressure flap actuator in an aircraft uses dynamic redundancy with cold standby *via* two independent DC motors acting on a single planetary gear.

FAULT TOLERANT COMMUNICATION

In-car networks used to interconnect electronics equipments becomes a key point for DBW systems. A common practice in the past was the extensive use of wiring. These wires significantly increased the complexity and weight of these systems. Today, in-car bus networks are largely adopted for the sake of reducing the vehicle weight and fuel consumption. The requirements of an automotive communication network are generated from the applications it has to support. Major automotive applications that involve networking and the usage of field buses are:

*   Drive By Wire
*   Chassis systems
*   Powertrain systems
*   Infotainment or multimedia

Two most important requirements for DBW systems are dependability and fault-containment. Distributed control is often used to meet these requirements. Distributed control algorithms require a coordinated snapshot of the controlled object. Such coordination can be achieved by the provision of a global time base.

Unification of bus systems within vehicles: Controller Area Network (CAN) will remain the selection of choice for event-triggered systems while a new protocol is needed for applications which require high performance, determinism, fault-tolerance, and flexibility. Several different field buses are used to address these various communication demands. A big issue that the automotive system producers deal with is that too many field bus technologies exist today. To interconnect these systems there is a need for high bandwidth together with flexibility and determinism. It is desirable to move from using too many technologies to fewer more general. To reduce the complexity evolving in a modern automotive system, it would be good to commit on a set of networking protocols that can be used in most of the applications typically.

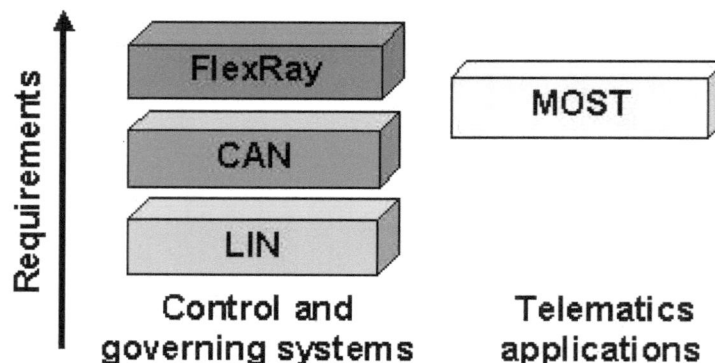

Fig. (8). Common field bus technologies between microprocessors.

The common field bus technologies interconnecting Electronic Control Units or ECUs today are: Controller Area Network (CAN), Local Interconnect Network (LIN), Byteflight, and Media Oriented Systems Transport (MOST). Fig. (**8**) shows a graphical illustration of the requirements for various communication protocols.

ANALYTICAL REDUNDANCY BASED FAULT TOLERANT CONTROL OF DBW SYSTEMS

Physical redundancy based fault tolerant control and communication protocols can significantly enhance the reliability. However, this methodology increases the overall system cost exponentially due to the presence of a large number of redundant hardware components. One possible solution to this problem is to replace the physical redundancy of the fault tolerant drive by wire systems by analytical redundancy. Analytical redundancy methodologies can not only reduce the overall system cost by reducing the total number of redundant components, but also further improve overall reliability of the system through the usage of a diverse array of sensory information. An up to date in-depth coverage of analytical redundancy based fault tolerant control and communication methodologies will be given in this section including predictive and nonlinear methodologies.

The concept of analytical redundancy has been investigated in the context of aerospace applications, primarily utilizing Eigen-structure theory. However, most of these articles were aimed at isolated subsystems in an aircraft or a spacecraft. Fly-By-Wire (FBW) systems are mostly based on full hardware redundancies (Bajpai *et al.*, 2001). As a result, analytical redundancy methodologies have not been utilized to a great extent in FBW systems. Nonetheless, the theoretical foundation for analytical redundancy methodology researched in aerospace applications can be useful in DBW system research. Introduction of Drive-By-Wire technology is more challenging in the automobile market, mainly because automobile consumers cannot afford the high cost of redundant systems the aerospace industry can. Each extra sensor, actuator, and Electronic Control Unit (ECU) increases the overall cost and weight of the vehicle. With the profit margin already low, this approach will not be acceptable to the automobile industries. By employing analytical redundancy techniques instead of hardware redundancy, it will be possible to bring the overall cost of such system down to a point that will be attractive to the automakers for mass production without sacrificing the high level of safety and reliability required by the consumers. Analytical redundancy along with appropriate fault detection, isolation, and accommodation methodologies can be utilized to make the DBW systems fault tolerant (Gadda *et al.*, 2007). Through analytical redundancy, the vehicle steering angle can be estimated from the states of other vehicle parameters without using an extra steering position sensor.

Sensor Analytical Redundancy in a Steer By Wire System

In this research, the analytical redundancy based fast fault detection algorithms was developed that based on physical models, nonlinear estimator and generalized predictive algorithms. In this algorithm, outputs from a number of redundant sensors as well as analytical sensor are checked against each other for a number of times before declaring a component to be faulty. The analytical sensor output is the combination exertion of the full vehicle model, nonlinear Sliding Mode Observer, and the long range prediction algorithm. The yaw rate signal can be measured with inexpensive sensors. Therefore with the measured yaw rate and the measured motor current input, the road wheel steering angles are estimated with the Sliding Mode Observer. Thereafter the steering angles were predicted by a long range predictor at variable prediction horizons with the participation of estimated steer angle and the motor current.

The proposed research concept is utilizes the long range prediction based fault detection, based on the analytical model of the SBW system (Fig. **9**) which would provide added safety to the SBW system *via* fast and robust fault detection and isolation of a component failure in such a system.

Since analytical redundancy methods are model-based, long-range prediction based Fault Detection, Isolation, and Accommodation (FDIA) algorithms are appropriate in such an application since modeling errors are inevitable in real-world systems. Furthermore, long-range prediction based FDIA methods (Fig. **10**) provide robustness against external disturbances which are expected in a DBW vehicle having a multitude of electric and electronics components. The fundamental concept in this proposition is that the sensor outputs are compared against the analytical counterpart (analytical redundancy) whose outputs are predicted several time step ahead *via* generalized predictive algorithm. In the event of a component failure, the predicted output will deviate from the sensor outputs several time steps ahead, thus reducing the detection latency. Table **1** lists the notations used in this section.

Table 1. Notations

θ	Road wheel angle
β	Vehicle body side slip angle
r	Vehicle yaw rate
a	Distance from the front tire to the vehicle's CoG
b	Distance from the rear tire to the vehicle's CoG
m	Vehicle mass
V	Vehicle longitudinal velocity
I_z	Vehicle moment of inertia
J_w	Moment of inertia of the road wheel
b_w	Viscous damping coefficient for wheel bearing
$C_{\alpha,f}$ and $C_{\alpha,r}$	Front and rear tire cornering coefficients
k_m	Motor torque constant
t_p	Pneumatic trail
t_m	Mechanical trail
τ_ϕ	Friction torque at the road wheel
i_m	Motor current
E_j and F_j	Uniquely defined polynomials in the Diophantine equation

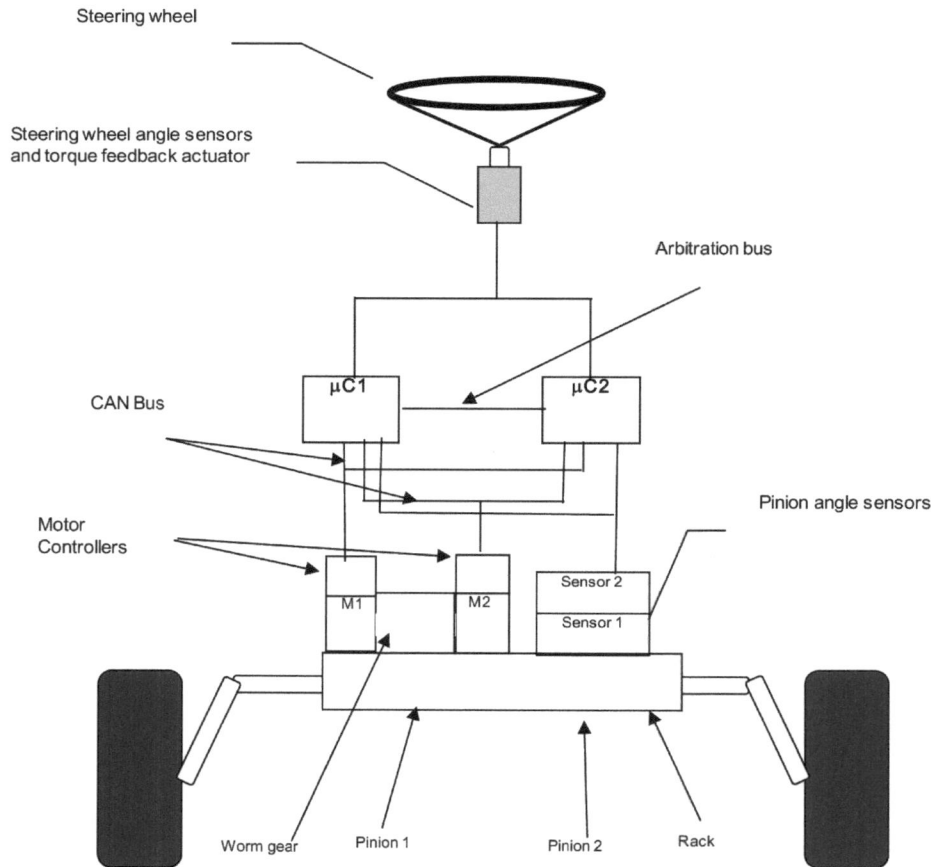

Fig. (9). Electronic architecture of an SBW system.

Observer and Predictor Based Modeling

An observer can be designed by combining the steering system model and vehicle model (Hasan & Anwar, 2008):

$$\dot{x} = Ax + Bi_m + E\tau_f$$

$$x = \begin{bmatrix} \beta & r & \theta & \dot{\theta} \end{bmatrix}$$

$$A = \begin{bmatrix} \dfrac{-C_{\alpha,f} - C_{\alpha,r}}{mV} & -1 + \dfrac{C_{\alpha,r}b - C_{\alpha,f}a}{mV^2} & \dfrac{C_{\alpha,f}}{mV} & 0 \\ \dfrac{C_{\alpha,r}b - C_{\alpha,f}a}{I_z} & \dfrac{-C_{\alpha,f}a^2 - C_{\alpha,r}b^2}{I_zV} & \dfrac{C_{\alpha,f}a}{I_z} & 0 \\ 0 & 0 & 0 & 1 \\ \dfrac{(t_p + t_m)C_{\alpha,f}}{J_w} & \dfrac{a(t_p + t_m)C_{\alpha,f}}{J_wV} & \dfrac{-(t_p + t_m)C_{\alpha,f}}{J_w} & \dfrac{-b_w}{J_w} \end{bmatrix}$$

$$B = \begin{bmatrix} 0 & 0 & 0 & \dfrac{k_m}{J_w} \end{bmatrix}^T \; ; E = \begin{bmatrix} 0 & 0 & 0 & -\dfrac{1}{J_w} \end{bmatrix}^T$$

(1)

$$C = \begin{bmatrix} 0 & 0 & 1 & 0 \end{bmatrix}$$

The motor current is the input to the system and the torque due to Coulomb friction is treated as a disturbance. The above system is fully observable.

Sliding Mode Observer (SMO)

The motivations for using Sliding Mode Observer are, it is model free and robust respect to bounded uncertainty. It can work under much less conservative condition. The idea underlying SMO observer design methods can be illustrated for a linear time-invariant system (Hasan & Anwar, 2008):

$$\dot{x} = Ax + Bu$$

(2)

$$y = Cx$$

$$y \in \Re^l, \; x \in \Re^n, \; rank\,(C) = l$$

The pair (C, A) is assumed to be observable and n is order of the system. A linear asymptotic observer is designed in the same form as the original system (2) with an additional input depending on the mismatch between the real values and the estimated values of the output vector:

$$\dot{\ddot{o}} = A\ddot{o} + Bu + L(y - C\ddot{o})$$

(3)

where $\ddot{o}$ is an estimate of the system state vector and $L \in \Re^{n \times l}$ is an input matrix. The state vector of the observer $\ddot{o}$ is available since the auxiliary dynamic system is implemented in a controller. The motion equation with respect to mismatch $\bar{x} = x - \ddot{o}$ is of form:

$$\dot{\bar{x}} = (A + LC)\bar{x}$$

(4)

The behavior of the mismatch governed by homogeneous Equation (4) is determined by eigenvalues of matrix (A+LC). For observable systems, they may be assigned arbitrarily by a proper choice of input matrix, L. It means that any desired rate of convergence of the mismatch to zero or estimate $\ddot{o}(t)$ to state vector x(t) may be provided. Then any full-state control algorithms with vector $\ddot{o}(t)$ are applicable.

The order of the observer may be reduced due to the fact that Rank (C) = 1 and the observed vector may be represented as:

$$y = C_1 x_1 + C_2 x_p$$
$$x = \begin{bmatrix} x_1 & x_p \end{bmatrix}$$
$$x_1 \in \mathfrak{R}^l, x_p \in \mathfrak{R}^{n-l}, \det(C_1) \neq 0$$

(5)

It is sufficient to design an observer only for vector xp, then the components of vector x_1 are calculated as,

$$x_1 = C_1^{-1}(y - C_2 x_p)$$

(6)

Write the system Equation (2) in space (y, xp) as,

$$\dot{y} = A_{11} y + A_{12} x_p + B_1 u$$
$$\dot{x}_p = A_{21} y + A_{22} x_p + B_2 u$$

(7)

where, $MAM^{-1} = \begin{bmatrix} A_{11} & A_{12} \\ A_{21} & A_{22} \end{bmatrix}, MB = \begin{bmatrix} B_1 \\ B_2 \end{bmatrix}, M = \begin{bmatrix} C_1 & C_2 \\ 0 & I_{n-l} \end{bmatrix}$

The coordinate transformation M in nonsingular, $\det(M) \neq 0$. Therefore, applying the simple Sliding Mode Observer (SMO) in the state space system equation,

$$\dot{\ddot{o}} = A\ddot{o} + Bu + L \, \mathrm{sgn}(y - C\ddot{o})$$

(8)

where, $\mathrm{sgn}(z) = col(\mathrm{sgn}(z_1), ..., \mathrm{sgn}(z_n))$

$$\mathrm{sgn}(z) = \begin{cases} +1 & if, z > 0 \\ -1 & if, z < 0 \end{cases}$$

Under a suitable choice of the gain matrix L in the observer, sliding occurs on the manifold $y - C\ddot{o} = 0$, and it becomes equivalent to the reduced order observer. The discontinuous vector function $v = L \, \mathrm{sgn}(y - \ddot{o})$. Now from Equation (7),

$$\dot{\ddot{o}} = A_{11} \ddot{o} + A_{12} \ddot{o}_p + B_1 u + L_1 \, \mathrm{sgn}(y - \ddot{o})$$
$$\dot{\ddot{o}}_p = A_{21} \ddot{o} + A_{22} \ddot{o}_p + B_2 u + L_2 \, \mathrm{sgn}(y - \ddot{o})$$

(9)

The system for the error $\bar{y} = y - \ddot{o}$ is of the form,

$$\dot{\bar{y}} = A_{11}\bar{y} + A_{12}\bar{x}_p + B_1 u + L_1 \, \mathrm{sgn}(\bar{y})$$

$$\dot{\bar{x}}_p = A_{21}\bar{y} + A_{22}\bar{x}_p + B_2 u + L_2 \, \mathrm{sgn}(\bar{y}) \tag{10}$$

The vector function $v \in \Re^l$ is chosen such that sliding mode is enforced in the manifold $\bar{y} = 0$ and the mismatch between the output vector y and its estimate $\dot{\hat{y}}$ reduced to zero. A vector L_2 must be found such that the mismatch $\bar{x}_p = x_p - \hat{x}_p$ between x_p and its estimate $\hat{x}_p$ decays as the desired rate. Equivalent value of the discontinuous function:

$$L_1 \, \mathrm{sgn}(\bar{y}) = A_{12}\bar{x}_p \tag{11}$$

For simplicity, L_1 is considered as 1 and equation (11) becomes:

$$\mathrm{sgn}(\bar{y}) = A_{12}\bar{x}_p \tag{12}$$

Now the equation on the sliding manifold appears from equation (10):

$$\dot{\bar{x}}_p = (A_{22} - L_2 A_{12})\bar{x}_p \tag{13}$$

Generalized Prediction Based Predictor

A class of predictive self-tuning controllers, known as Generalized Predictive Controller (GPC) (Hasan and Anwar, 2008) have shown robustness against unstable plants, non-minimum-phase plants, model over parameterization, and uncertain process dead time. These controllers have also been observed to provide offset free behavior for the closed loop system since they include an integral action. These set of controllers have been very successful in regulator or tracking type observer applications. In the context of long range prediction, the prediction horizon, j is a tunable design variable that can be set to any value according to the desired prediction range. The predictive nature of the GP (Generalized Prediction) based predictor algorithm comes from the use of the Diophantine equation. Through the use of Diophantine equation, the output of the plant is predicted j-step ahead of present time. This prediction output is then used for future fault detection and identification.

Considering the state space equation (2), a transfer function can be obtained as follows:

$$G(s) = \frac{U(s)}{Y(s)} = C(sI - A)^{-1} B \tag{14}$$

After discretization of the equation (14), a more general form of the dynamic model of the dynamic model of the vehicle model can be written as:

$$A(z^{-1})Y(z) = B(z_{-1})U(z) \tag{15}$$

where $A(z^{-1})$ and $B(z^{-1})$ are polynomial of order n_a and n_b respectively in the backward shift operator in time, z^{-1}. $A(z^{-1})$ and $B(z^{-1})$ have the following forms:

$$A(z^{-1}) = d_0 + d_1 z^{-1} + d_2 z^{-2} + \ldots + d_{na} z^{-na}$$
$$B(z^{-1}) = n_0 + n_1 z^{-1} + n_2 z^{-2} + \ldots + n_{nb} z^{-nb} \tag{16}$$

In order to make a prediction of the future output of the road wheel angle, the Diophantine identity is used to derive the j-step ahead prediction of $\Delta U(t+j)$.

$$1 = E_j(z^{-1})A(z^{-1}) + z^{-j}F_j(z^{-1}) \tag{17}$$

Where E_j and F_j are uniquely defined polynomials for a given $A(z^{-1})$ and the prediction interval j. In the present work, the recursive technique has been used to obtain E_j and F_j (Hasan & Anwar, 2008). This makes the procedure computationally very efficient. It has been shown that with increasing j only the highest order term in $E_{j+1}(z^{-1})$ changes while the rest of the coefficients remain the same in $E_j(z^{-1})$. Therefore, we can write:

$$E_{j+1}(z^{-1}) = E_j(z^{-1}) + e_j z^{-j} \tag{18}$$

where, $E_j(z^{-1}) = e_0 + e_1 z^{-1} + e_2 z^{-2} + ... + e_{j-1} z^{-(j-1)}$

In the degree of polynomial $A(z^{-1})$ is n_a, then the degree of $F_j(z^{-1})$ becomes n_a. The coefficients of the polynomial $F_j(z^{-1})$ may then be denoted as:

$$F_j(z^{-1}) = f_{j,0} + f_{j,1} z^{-1} + f_{j,2} z^{-2} + ... + f_{j,na} z^{-na} \tag{19}$$

Steering Angle Prediction

GP based prediction supposed to be executed for the discrete model. Hence the forth order vehicle model has been discretized to make itself compliant with the predictor. From the equations (15) – (16), the Diophantine prediction equation (j-step ahead predictor) is given by,

$$E_j(z^{-1})(d_0 + d_1 z^{-1} + d_2 z^{-2} + d_3 z^{-3} + d_4 z^{-4})\Delta + z^{-j}F_j(z^{-1}) = 1 \tag{20}$$

Multiplying equation (20) with $\Box(t+j)$ and rearranging that equation, we obtain:

$$\theta(t+j) = F_j(z^{-1})\theta(t) + E_j(z^{-1})(n_0 + n_1 z^{-1} +$$
$$n_2 z^{-2} + n_2 z^{-3} + n_4 z^{-4})\Delta i_m(t-j+1) \tag{21}$$
$$where, \Delta = (1 - z^{-1})$$

Equation (21) predicts the value of the pinion angle θ in the future (j - time step ahead).

$$\theta(t+j) = F \times \theta(t) + EB \times \Delta i_m(t-j+1) \tag{22}$$

The matrices $F \in \Re^{N \times 5}$ and $EB \in \Re^{N \times (N+5)}$ are calculated by using the MATLAB script.

HARDWARE-IN-LOOP (HIL) BENCH SIMULATION RESULTS

The build of the hardware-in-loop (HIL) bench for the SBW system involves two steps, namely, control hardware and mechanism construction. A parts list for both control hardware and mechanism construction is shown in Table **2**.

Table 2. List of Components/Parts for Building the Loop Bench

Name of Component/Part	Quantity	Description
Servo Motor	1	Parker BE342KJ-K10N, with a gear box coupled with the rack.
Drive	1	Parker OEM770T, selected for the servo motor.
Rack	1	Volkswagen SI-9281-9-2, with a pinion on top
Angular Sensor	3	Model: MH22B, coupled with the pinion
Steering Wheel System	1	The system has been already built before, with angular signal as output.
Controller	1	DSpace MicroAutoBox 1401

Figs. (**11a**) and (**11b**) show the SBW HIL bench mechanical and control hardware.

Initialization of Vehicle Model for Simulation

As illustrated in Fig. (**10**), a full vehicle model is used for the HIL bench. The vehicle model was based on data for a compact vehicle with the following parameter initialization (Anwar & Chen, 2007):

$Fw = 2$ (N-m); $Jw = 0.5$ (kg-m^2); $bw = 20$; $km = 0.5$; $I = 4000$ (kg-m^2); $Cf = 22000$ (N/rad); $Cr = 55000$ (N/rad); $m = 1400$ kg; $V = 15$ (m/s); $a = 1.0$ (m); $b = 1.5$ (m); $C_0 = (Cf + Cr)$ (N/rad); $C1 = (bCr - aCf)$ (N-m/rad); $C2 = (a^2Cf + b^2Cr)$ (N-m^2/rad); $tp = 0.025$ (m); $tm = 0.03$ (m); $C_3 = (tp + tm)Cf$ (N-m/rad); $K = 80$; $T = 0.005$ (sec).

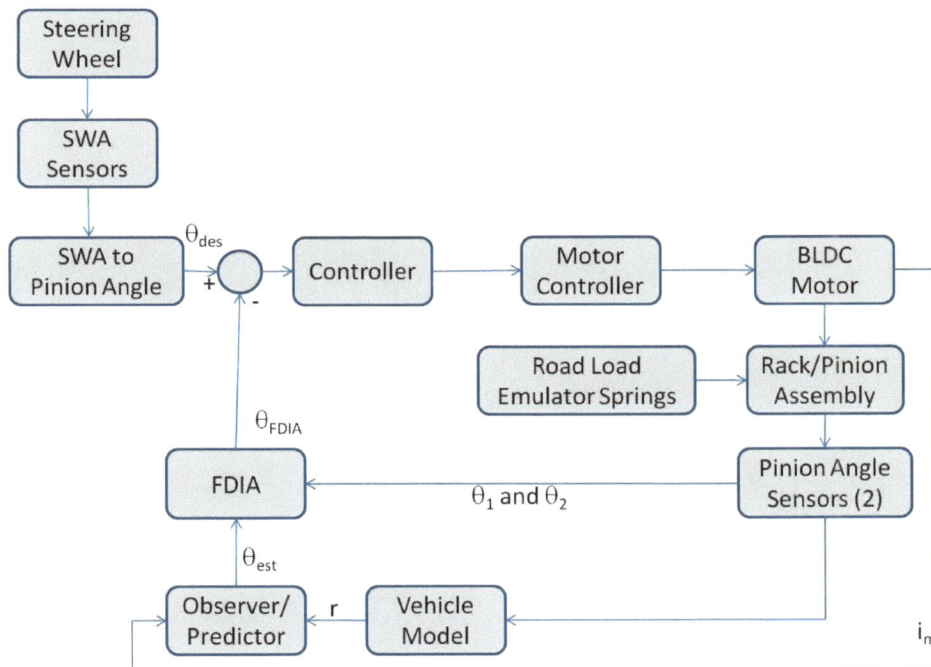

Fig. (10). Road wheel angle estimation and fault detection, isolation, and accommodation (FDIA) on the Hardware-In-Loop (HIL) bench.

The vehicle model presented in this thesis is evaluated on a validated SIMULINK model (Anwar & Chen, 2007). The cornering coefficients have been considered for a light weight passenger Front Wheel Drive (FWD) car. Therefore the mass and dimensions of the vehicle were perceptibly standard for the small passenger car. The simulation process has been appraised for a slow moving vehicle and the pneumatic and mechanical trails have been carried out from the standard passenger car tires specification. The driver's factor is chosen as nominal driving effect. But the sample time can be deviated according to the dynamic requirement of the system.

Faults Types and Their Implications

Two major fault types have been introduced into a steer by wire system, namely:

• Permanent fault

• Incipient fault

The persistent fault type is illustrated in Fig. (**12a**). Amplitude change fault can be either positive amplitude change or negative amplitude change types. Incipient fault should be handled at the early stage of the system operation otherwise according to their nature; they are gradually increased to a larger extent that could be difficult to control. The incipient faults with attenuating amplitude are shown in Fig. (**12b**).

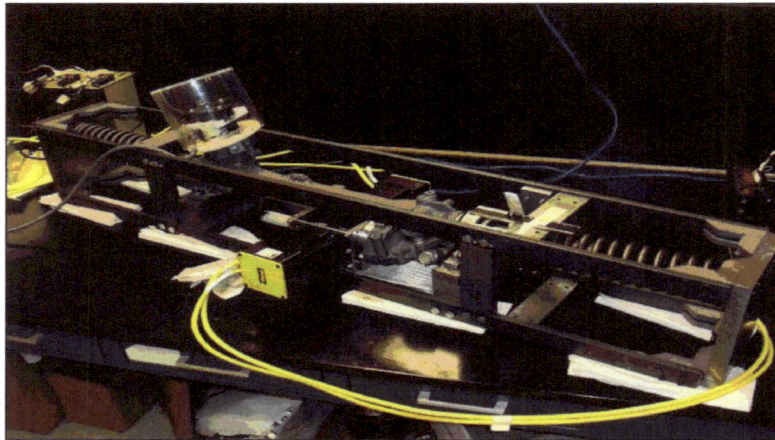

Fig. (11a). Completed construction of the hardware.

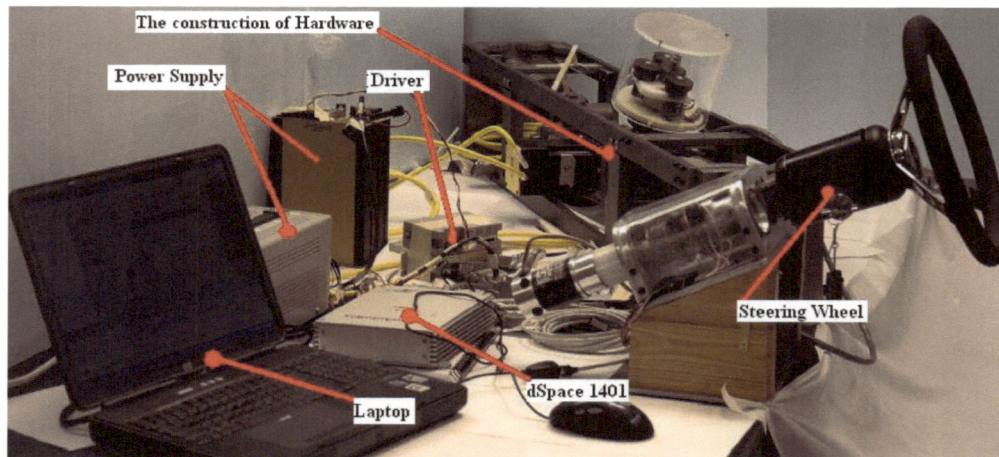

Fig. (11b). Complete SBW hardware with all hardware components.

These two common types of faults are introduced into the vehicle model system to verify the FDIA methodology along with SMO and GP based predictor for the SBW system.

In order for the SBW system to be robust, the sensor measurements must be accurate and reliable. Therefore, any faulty signal must be eliminated to prevent undesirable steering effects. The Fault Detection, Isolation, and Accommodation algorithm (FDIA) (Anwar & Chen, 2007) used in this paper is able to handle single point faults without interrupting the functionality of the SBW system. This algorithm can be easily modified to handle multiple faults if more than three-sensor signals are compared. The FDIA algorithm implemented in SIMULINK below is based on a majority voting scheme in which a minimum of three signals are required for this scheme to work. The sensor signals are compared against each other in real-time to determine the faulty signal where majority is assumed to be correct. This is based on the assumption that the event of multiple simultaneous sensor failures is very rare. This algorithm can determine which sensor has failed by comparing its value against other sensors' values. This algorithm can manage hard-failures as well as soft-failures. Hard-failure is characterized by an abrupt or sudden sensor failure and soft-failure is characterized by biases or drifts in the signal over time. When a sensor fails, its signal is no longer used in the road wheel angle calculation. In such a situation the driver would be alerted of the sensor failure, but would also still be able to maintain safe control of the vehicle.

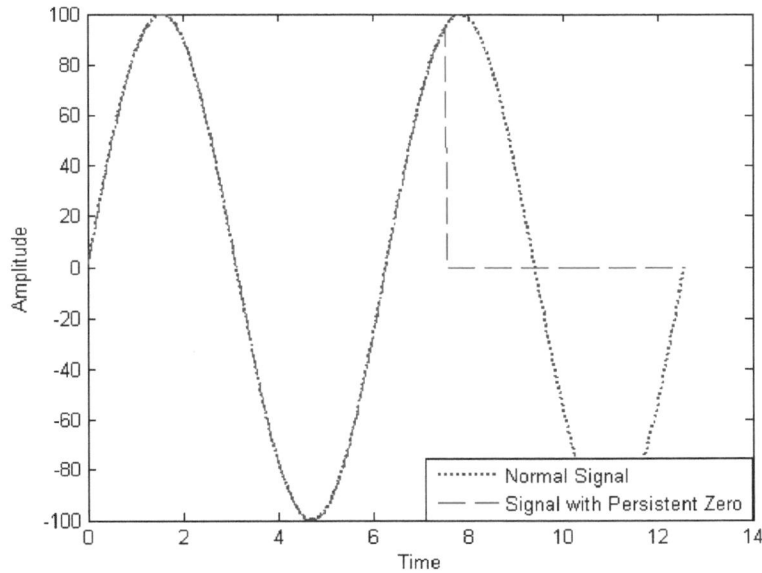

Fig. (12a). Persistent zero sensor fault introduced in one of the two physical sensors.

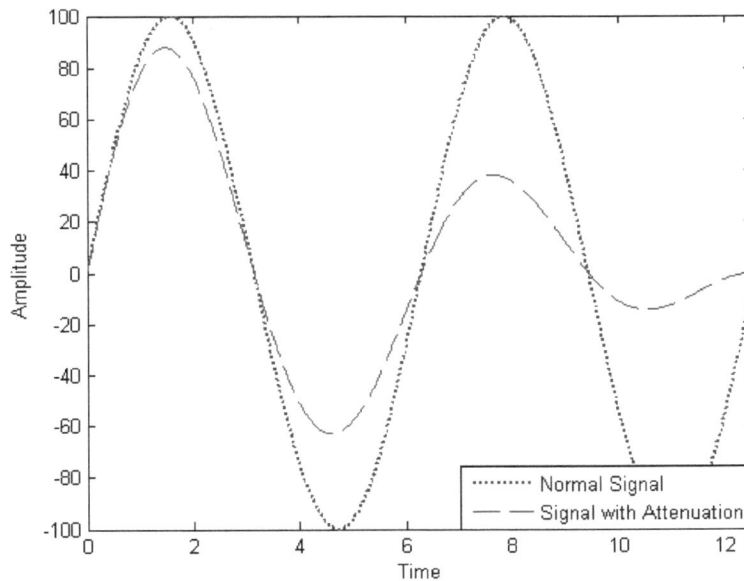

Fig. (12b). Sensor fault with attenuating amplitude in one of the two physical sensors.

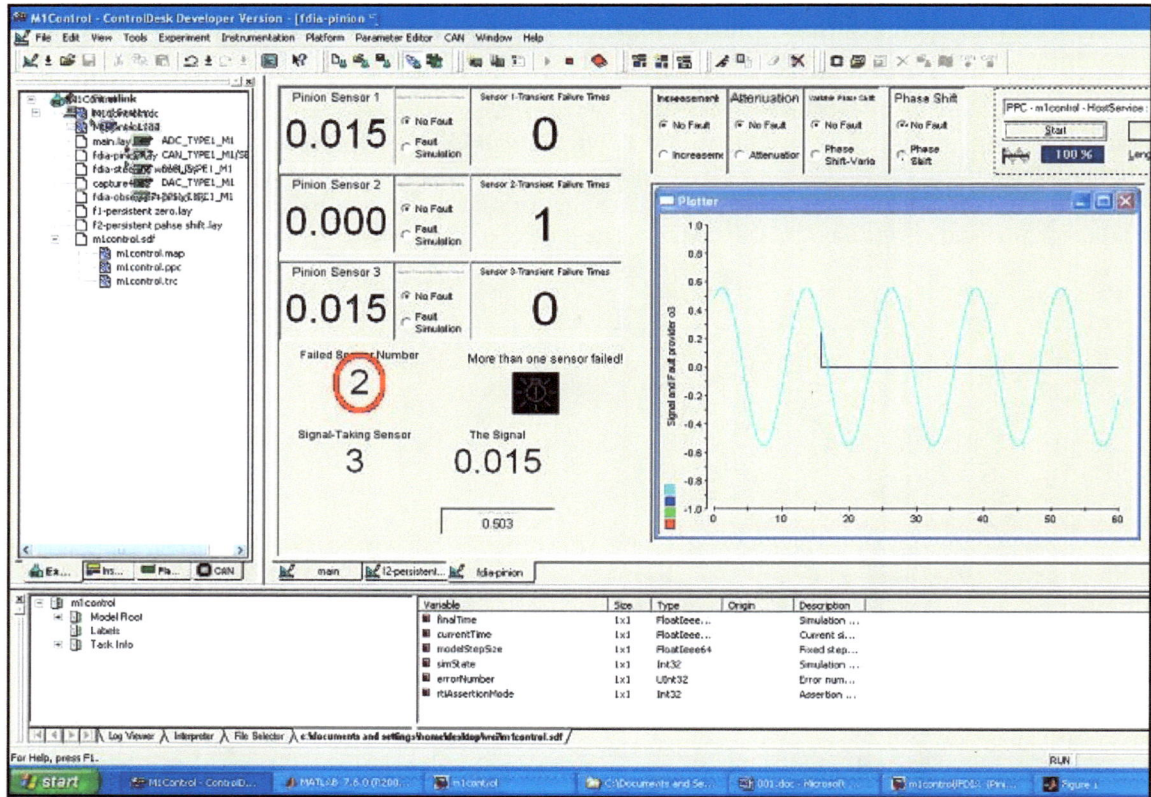

Fig. (13a). Persistent zero type fault introduced in one of the two physical sensors with the fault state circled on dSPACE ControlDesk GUI in real-time.

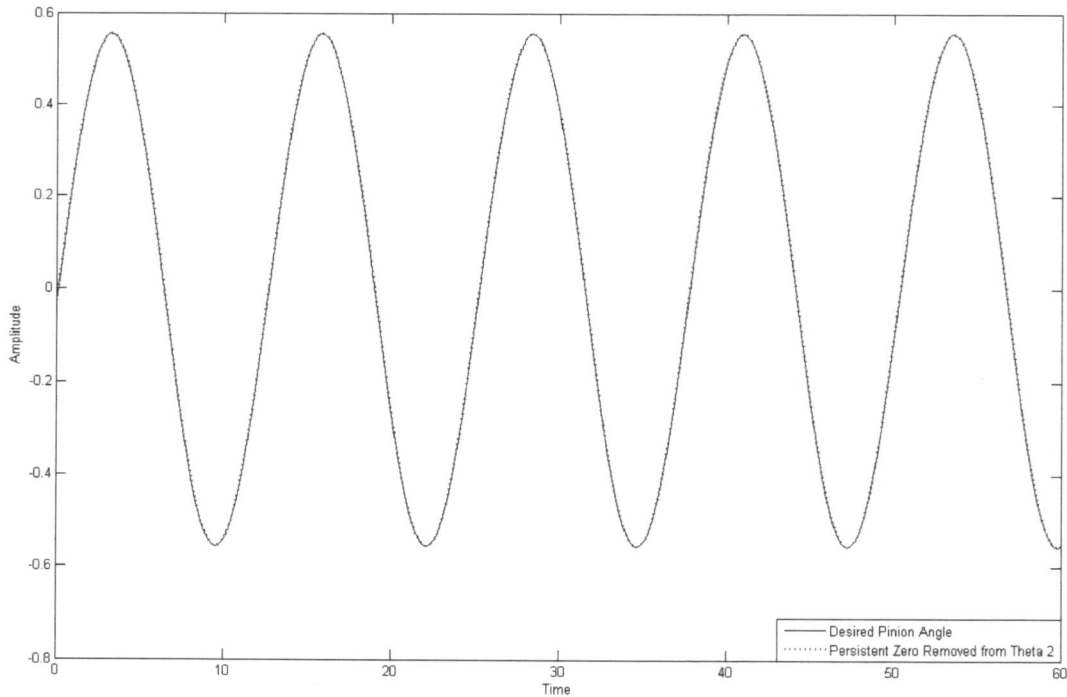

Fig. (13b). FDIA output of sensor data after removing the persistent zero type fault in one physical sensor.

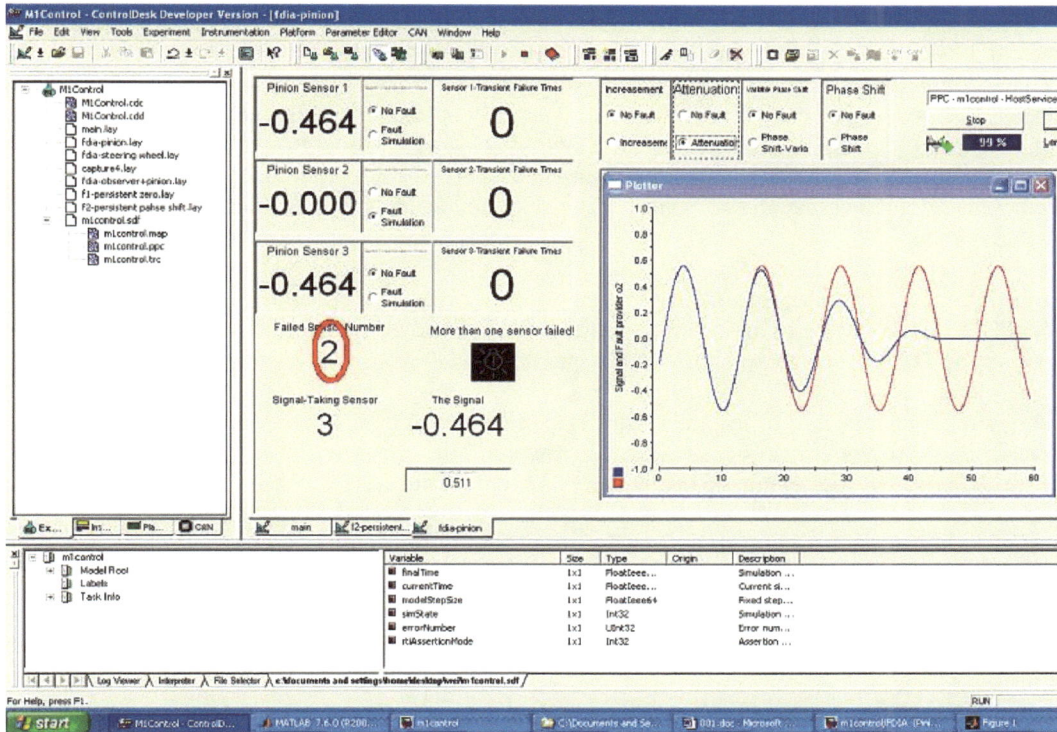

Fig. (14a). Attenuating amplitude type incipient fault introduced in one of the two physical sensors with the fault state circled on dSPACE ControlDesk GUI in real-time.

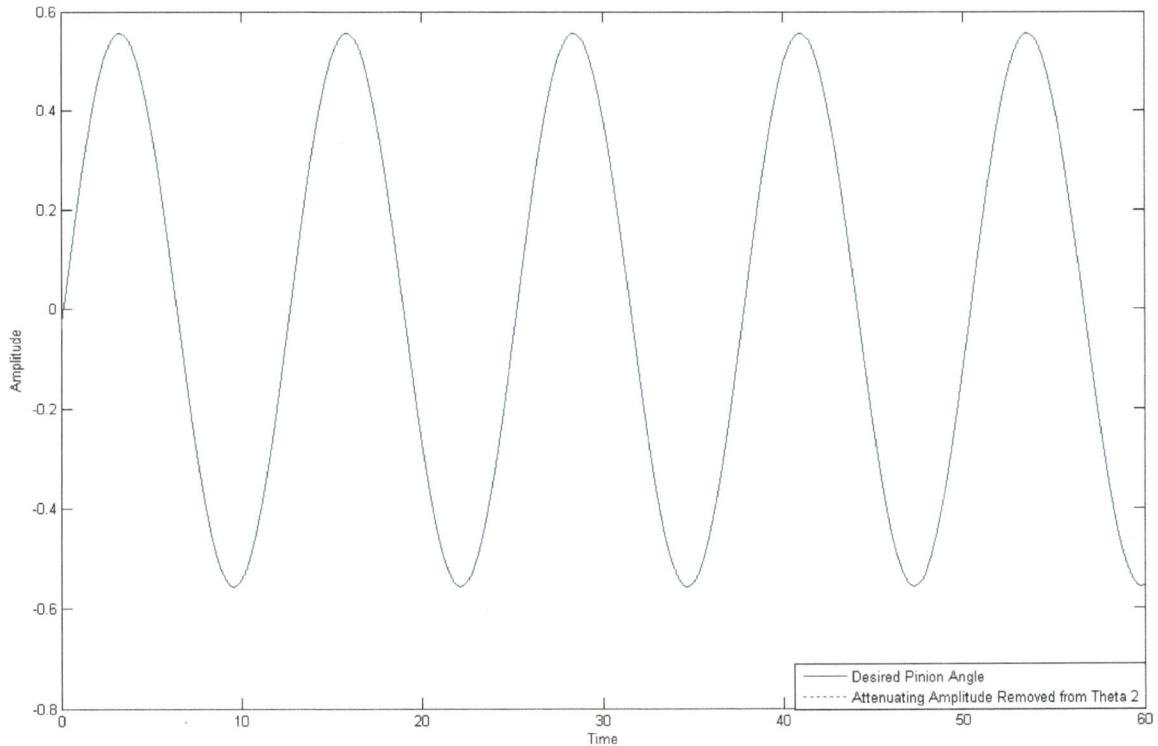

Fig. (14b). FDIA output of sensor data after removing the attenuating amplitude type incipient fault in one physical sensor.

A number of experimental simulation runs were performed on the HIL bench in order to evaluate the developed methodology of estimating and then predicting the road wheel angle to detect and control them. And some simulations were observed to verify the advantages of higher prediction horizon into the dynamic systems. The SBW controller, the yaw angle observer, the road wheel angle estimator, the FDIA algorithms, and the Generalized Predictive (GP) based predictor are combined with a simplified vehicle model with an SBW actuation system. The combined model was given a sinusoidal steering input. Fault was then injected to one of the three road wheel angle sensors whether one of them was analytical sensor. Fault flags and the output of the road wheel angle from the FDIA block are then recorded.

Persistent zero introduce fault was injected by making the sensor out a constant value. It was considered a faulty signal by the FDIA block and was eliminated from the output as shown on the real-time software GUI (Fig. **13a**). Fig. (**13b**) shows the FDIA block output with the removal of the fault from the system.

Fig. (**14a**) shows the sensor signals of the attenuating amplitude type incipient faults on a ControlDesk GUI capture. This fault is introduced to the 2nd physical sensor *via* software. This shows the incipient fault as gradually changes with time. The faulty signal was eliminated from the FDIA output signal (Fig. **14b**) validating the effectiveness of the GP based predictive estimation of the proposed algorithms.

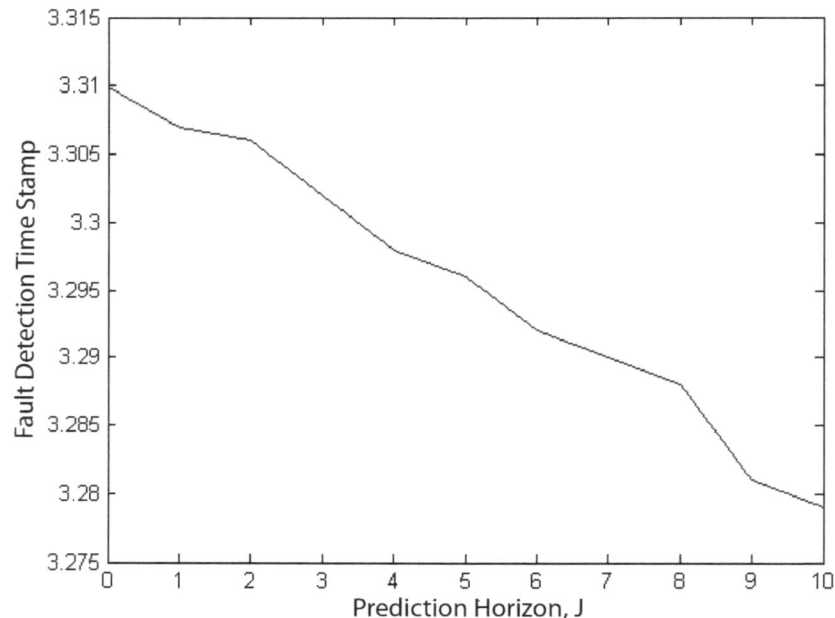

Fig. (15). Effect of prediction horizon on fault detection speed for the attenuating amplitude type incipient fault.

The effect of GP based prediction with the different prediction horizons on fault detection speed is shown in Fig. (**15**). The effect of prediction horizon for the various magnitudes of attenuating amplitude type incipient fault is illustrated by the fault detection time which decreases significantly with increase in the prediction horizon. Thus the overall efficiency of fault detection is improved by incrementing the prediction horizon in the proposed GP based estimation.

It was demonstrated that it is possible to increase the level of robustness of a fault tolerant SBW system *via* successful implementation of SMO and GP based fault tolerant control. In the present work, a nonlinear Sliding Mode Observer (SMO) was designed and implemented to estimate the road wheel angle with available sensor output of yaw angle and motor input current. Through the proposed predictive analytical redundancy based fault detection and isolation algorithm, an extra level of redundancy was possible without any extra hardware. The proposed algorithms rendered the SBW system with a robust fault tolerant system as evidenced by the simulation results. It was also observed that the reliability of the proposed methodology increases with the increase of prediction horizon as it reduces the detection time for the faulty sensors.

FUTURE TRENDS

Various drive by wire systems in the same vehicle may share a diversity of sensors and actuators in order to bring down the overall system cost. This poses a unique challenge/opportunity for integration of all the sensory information *via* data fusion methodologies that will take the analytical redundancy features of these systems to a new level. In addition, integration of fault diagnosis and control bring about new challenges since the control objectives and diagnosis objectives may have conflicting features. For example, in an SBW system, the control objectives could be noise/disturbance rejection and command tracking whereas the diagnosis objective could be faulty signal tracking and noise reduction (Zheng *et al.*, 2006). Integrated control of various drive by wire systems (e.g. TBW, BBW, SBW) is also being investigated for future application on DBW vehicles (Abe & Mokhiamar, 2007; Sameshima *et al.*, 2004; Lu & Hedrick, 2005; Setlur *et al.*, 2006). Robust sliding mode control of a throttle body for drive by wire operation of automotive engines has been critical in the success of TBW systems (Rossi *et al.*, 2000).

For safety-critical systems, communication reliability cannot be based on probabilistic measures. A CAN system is not possible to calculate the communication delay except for some high priority message. FlexRay guarantees the transport for safety-critical message with known latency. FlexRay offers deterministic system design. FlexRay combines a time-triggered along with an event-triggered system. Based on an extended TDMA (time-division multiple access) media access strategy, a so-called communication cycle is set up, consisting of a mandatory static segment, an optional dynamic segment and two protocol segments called symbol window (optional) and network idle time (mandatory). Application cycles can be formed by one or more communication cycles, which are executed from start-up of the network until its shutdown. A time-triggered communication system depends upon a common time base, the so-called global time. This time base has to be shared by all communication controllers in the network. In order to synchronize the clock, certain messages are marked as synchronization messages in the static segment. By means of a special algorithm, the local clock-time of a component, for example, a communication controller, is corrected in such a way that all local clocks run synchronously to a global clock.

A bus guardian ensures that the data traffic moves in a timely fashion. Access to the communication bus has to be carefully managed, especially with regard to safety relevant applications. An independent component that protects a channel from interference is needed. It must limit the times that any communication controller can transmit to those pre-assigned intervals, in which it is allowed to do so. This component is the Bus Guardian. Basically, FlexRay systems allow to use bus guardians with independent clocks and are configured in a way that an error in the controller cannot influence the guardian and vice versa.

CONCLUDING REMARKS

It is clear that drive by wire systems have many benefits over their conventional counterpart in the area of safety and performance. DBW systems can, in fact, enhance the safety and performance of a vehicle equipped without any additional hardware, since these enhancements can be achieved through software upgrades. While drive by wire systems have made significant progress in fault tolerance and reliability, these systems, particularly the safety critical BBW and SBW systems, are still some steps away from full commercialization. One of the main reasons for this is the high cost of these systems due to multiple redundancies in sensors, actuators, microcontroller, and communications buses. As demonstrated in this chapter, analytical redundancy/model based fault tolerant control can reduce the total number of redundant components to certain extent. Further improvement in vehicle safety and reliability as well as additional cost reduction can be achieved *via* innovative actuator design, manufacturing process optimization for microprocessors and communications buses, and inexpensive sensor development with diversity in the sensors. The timing of DBW commercialization will largely depend on further research and development of cost effective fault tolerant systems in order to address the aforementioned challenges and the willingness of the automakers and their suppliers to take on such a task to mass produce DBW systems.

REFERENCES

Abe, M., & Mokhiamar, O. (2007). An integration of vehicle motion controls for full drive-by-wire vehicle. Proc. IMechE Part K: *Journal of Multi-body Dynamics, 221*, 117-127.

Anwar, S. (2005). Generalized Predictive Control of Yaw Dynamics of a Hybrid Brake-By-Wire Equipped Vehicle. *International Journal of Mechatronics, 15*, 1089-1108.

Anwar, S. (2006). Anti-Lock Braking Control of a Hybrid Brake-By-Wire System. *Journal of Automobile Engineering – IMechE Proceedings Part D, 220*(8), 1101-1117.

Anwar, S. & Chen, L. (2007). An Analytical Redundancy Based Fault Detection and Isolation Algorithm for a Road-Wheel Control Subsystem in a Steer-By-Wire System. IEEE Transactions on Vehicular Technology, 56(5), 2859-2869.

Bajpai, G., Chang, B.C., & Lau, A. (2001, September). Reconfiguration of Flight Control Systems for Actuator Failures. *IEEE AESS Systems Magazine*, 29-33.

Chang, S.C. (2007). Adoption of state feedback to control dynamics of a vehicle with a steer-by-wire system. *Proc. IMechE Part D: J. Automobile Engineering, 221*, 1-12.

Gadda, C.D., Laws, S.M., & Gerdes, J.C. (2007). Generating Diagnostic Residuals for Steer-by-Wire Vehicles. *IEEE Trans. on Control Sys. Tech., 15*(3), 529-540.

Haggag, S., Alstrom, D., Cetinkunt, S., & Egelja, A. (2005). Modeling, Control, and Validation of an Electro-Hydraulic Steer-by-Wire System for Articulated Vehicle Applications. *ASME/IEEE Trans. on Mechatronics, 10*(6), 688-692.

Hasan, M.S. & Anwar, S. (2008). Sliding Mode Observer Based Predictive Fault Diagnosis of a Steer-By-Wire System. Proceedings of the 17th Int'l Federation of Automatic Control (IFAC) World Congress, Seoul, S. Korea.

Hebden, R.G., Edwards, C., & Spurgeon, S.K. (2004). Automotive steering control in a split-mu maneuver using an observer based sliding mode controller. *Vehicle System Dynamics, 41*(3), 181-202.

Higgins, A. & Koucky, S. (2002, May 9). Mercedes pumps 'Fly By Wire' brakes into new roadster. Machine Design, 26.

Hoseinnezhad, R. (2006). Position Sensing in Brake-By-Wire Calipers Using Resolvers. *IEEE Transactions on Vehicular Technology, 55*(3), 924-932.

Iserman, R. (2006). Fault-Diagnosis Systems - An Introduction from Fault Detection to Fault Tolerance. Heidelberg (Germany): Springer.

Isermann, R. (2005). Mechatronic Systems - Fundamentals. Heidelberg (Germany): Springer.

Isermann, R. & Münchhof, M. (2011). Identification of Dynamic Systems - An Introduction with Applications. Heidelberg (Germany): Springer.

Isermann, R., Schwarz, R., & Stolzl, S. (2002, October). Fault tolerant drive-by-wire systems. IEEE Control Systems Magazine, 64-81.

Limpibunterng, T. & Fujioka, T. (2002). A new design approach for steer-by-wire system by dual port system. *Vehicle System Dynamics*, Suppl. 37, 197-208.

Limpibunterng, T. & Fujioka, T. (2004). Bilateral driver model for steer-by-wire controller design. *Vehicle System Dynamics*, Suppl. 41, 381-390.

Lu, X-Y. & Hedrick, J. K. (2005). Heavy-duty vehicle modelling and longitudinal control. Vehicle System Dynamics, 43(9), 653–669.

Münchhof, M., Beck, M., & Isermann, R. (2009). Fault Diagnosis and Fault Tolerance of Drive Systems: Status and Research. *European Journal of Control, 15*(3-4), 370-388.

Münchhof, M., Beck, M., & Iserman, R. (2009). Fault Tolerant Actuators and Drives - Structures, Fault Detection Principles and Applications. Proceedings of the 7th IFAC International Symposium SAFE Process, Barcelona, Spain.

Oh, S-W., Chae, H-C., Yun, S-C., & Han, C-S. (2004). The design of a controller for the steer-by-wire system. *JSME International Journal, Series C, 47*(3), 896-907.

Rossi, C., Tilli, A., & Tonielli, A. (2000). Robust Control of a Throttle Body for Drive by Wire Operation of Automotive Engines. *IEEE Transactions on Control System Technology, 8*(6), 993-1002.

Sameshima, H., Ogasa, M., & Yamamoto, T. (2004). On-Board Characteristics of Lithium Ion Batteries for improving energy regenerative efficiency. *Quarterly report of RTRI, 45*(2), 45-52.

Segawa, M., Kimura, S., Kada, T., & Nakano, S. (2004). A study of the relationship between vehicle behavior and the steering wheel torque on steer by wire vehicles. *Vehicle System Dynamics*, Suppl. 41, 202-211.

Setlur, P.J., Wagner, R., Dawson, D.M., & Braganza, D. (2006). A Trajectory Tracking Steer-by-Wire Control System for Ground Vehicles. *IEEE Transactions on Vehicular Technology, 55*(1), 76-85.

Stanton, N. & Marsden, P. (1997). Drive-by-wire systems: some reflections on the trend to automate the driver role. *Proceeding of Inst. Mech. Engrs. Part D, 211*, 267-276.

Yih, P. & Gerdes, J.C. (2005). Modification of Vehicle Handling Characteristics *via* Steer-by-Wire. *IEEE Trans. on Control Sys. Tech, 13*(6), 965-976.

Zheng, B. & Anwar, S. (2008). Fault Tolerant Control of the Road Wheel Subsystem in a Steer-By-Wire System. *International Journal of Vehicular Technology*, Article ID 859571, doi:10.1155/2008/859571, 1-8.

Zheng, B., Huang, M., Daugherty, B., & Anwar, S. (2006). Integrated Steer-By-Wire control and diagnostic system. Proceedings of the American Control Conference, Minneapolis, Minnesota.

CHAPTER 2

System Level Reliability Issues and Their Enhancement in Drive-by-Wire (DBW) Systems

M. Abul Masrur *

US Army RDECOM-TARDEC, RDTA-RS, MS-233, 6501 E. 11 Mile Road, Warren, MI 48397-5000, USA

Abstract: Drive-by-wire systems in automobiles and fly-by-wire systems in aircrafts have made these vehicles versatile with added features and benefits, along with ease of reconfiguration, in particular during graceful degradation. However, there are certain consequences of such by-wire systems in terms of overall system reliability. The issue has not been quantitatively discussed in the literature in a systematic manner, to the best of the knowledge of the author. The chapter discusses details of the drive-by-wire system architectures from a system viewpoint, analyzes the reliability with quantitative metric, and indicates methods of enhancing reliability by using both hardware and software redundancies.

Keywords: Drive By Wire, Fly By Wire, Multiplexed Systems, System Load Availability, Multi-Protocol Communication, Controller Area Network (CAN).

SYSTEM VIEWPOINT

Drive-by-wire (DBW) for automobiles (Isermann *et al.*, 2002) or its counterpart fly-by-wire (FBW) (Seidel, 2009), and in general X-by-Wire, and sometimes termed "multiplexed system" in a general sense, have significantly helped to achieve better size, weight, packaging, and overall performance in ground vehicles (including airborne and waterborne vehicles, i.e. aircrafts, ships etc.), allowed easier inclusion of higher number of components and subsystems within the vehicle in a modular fashion, allowed multiple platforms using common architecture, and contributed to overall flexibility in manufacture. Multiplex system "normally" involves computer type of network, where several signals can be combined (or "multiplexed") and sent out through the same common media like copper wire, fiber optic cables etc. Although "by-wire" systems these days involve communication networks, in principle it need not be so. In other words, it is possible to use direct wirings to actuate loads separately. These topics have been discussed in other sections within this eBook. However, incorporation of multiplexing has also introduced certain risk factors as well, in terms of catastrophic failure conditions. Since DBW incorporates various components and subsystems, and since its successful functionality leads to the overall system performance, it is worthwhile to look at it from a holistic and system level viewpoint. The purpose of this section will be to discuss DBW from a total system level perspective to see what is involved to enhance its reliability and how component and subsystem level reliability can affect the overall system reliability. We will also try to quantify the overall system reliability with numerical example. With the above in view, it is necessary to study the overall architecture of the system first.

TYPES OF THINGS DEALT WITH IN DBW

As is well known, DBW (or equivalently FBW etc) is intended to "make something happen" or actuate something (a device or "load") at a distance by communicating from remote, and in consequence, the source of power or energy

---

*Address correspondence to M. Abul Masrur: US Army RDECOM-TARDEC, RDTA-RS, MS-233, 6501 E. 11 Mile Road, Warren, MI 48397-5000, USA; E-mail: md.abul.masrur@us.army.mil

can be at a different location than the actuation requesting agent or the load. Besides actuation of some component, multiplexing can also imply flow of signals in response to some request, e.g. speed of a vehicle, where a transducer signal due to the speed will be taken to the instrument cluster through some signal wire, and not through the actual transducer component, and thus it is not necessarily that only high power items are commanded through multiplex system. These are the main categories of DBW. FBW is no different. Obviously, then, multiplexing involves both hardware and software (software is involved in communication network and also actuation algorithm etc.) and both are equally important for its successful operation.

ACTUATION PROCESS IN DBW

As we noted above, it may be our interest to actuate something at a distance through DBW. For example, we might need to steer the vehicle wheel, or move the aircraft flaps in the wing, or the rudder of a ship. The final actuation may be implemented by means of a hydraulic device, e.g. a hydraulic pump driving a hydraulic motor or perhaps a hydraulically driven piston etc., or it could be an electric motor to do the same. In the former case, the source of hydraulic energy could be far away from the requester and located more conveniently near the load, with all the hydraulic plumbing, pipes etc. packaged conveniently at a suitable location. In the latter case if the actuation is done electrically, and the battery or the electrical power source could be located at a convenient place and the actuation motor could be very next to the load. Of course, besides hydraulic or electric it is possible, like earlier days, to have direct mechanical linkage (through linkages, levers etc.) from the driver which will actuate something – though in a very cumbersome way. Similarly "non-power" items like speed or odometer signal etc. can be communicated from the source to the instrument cluster through direct link, or through hydraulic link, or direct electrical wirings.

ENERGY SOURCE AND ACTUATORS

It is obvious, therefore, from the above that we either need hydraulic, electric, or even direct mechanical linkage to actuate something or receive signals. Multiplexing (or DBW, FBW) has taken out that "directness" between two points through communication network.

ARCHITECTURE

A system level architecture of a multiplex system is shown below (Masrur *et al.*, 2004). In this system, the red line indicates the power bus and the blue line indicates the communication signal line, which in the case of a CAN (controller area network) network can be a twisted pair copper wire. The benefits of the above are obvious. The power line can be simple with less connectors etc. The information from the driver does not need a high power cable to activate the power switch, only low power signal wire (thin in size) is sufficient. The rectangular boxes are what is known as ECU (electronic control unit) or nodes. Each node can cater several loads (the oval shaped ones). The information in the "blue" line can operate in a multiplexed manner, i.e. several ECU's can share the same wire using some protocol or communication "mannerism". For CAN protocol, there are priorities (Masrur, 1989) based on which a particular function will get hold of the bus. For safety critical devices there can be different protocols, e.g. TTP (time triggered protocol), Flexray, or sometimes there can be mixed protocols (Arora *et al.*, 2004) and it is possible to have separate and dedicated bus to deal with safety critical loads and to have some kind of gateway between multiple communication buses, somewhat like a computer network or internet gateways.

WHAT CAN HAPPEN DURING ECU FAILURE

Obviously, it can be seen from the above architecture in Fig. (1) that if an ECU fails, it will take down with it several loads which won't operate any more. That is a very dangerous situation, especially if the loads are vital ones.

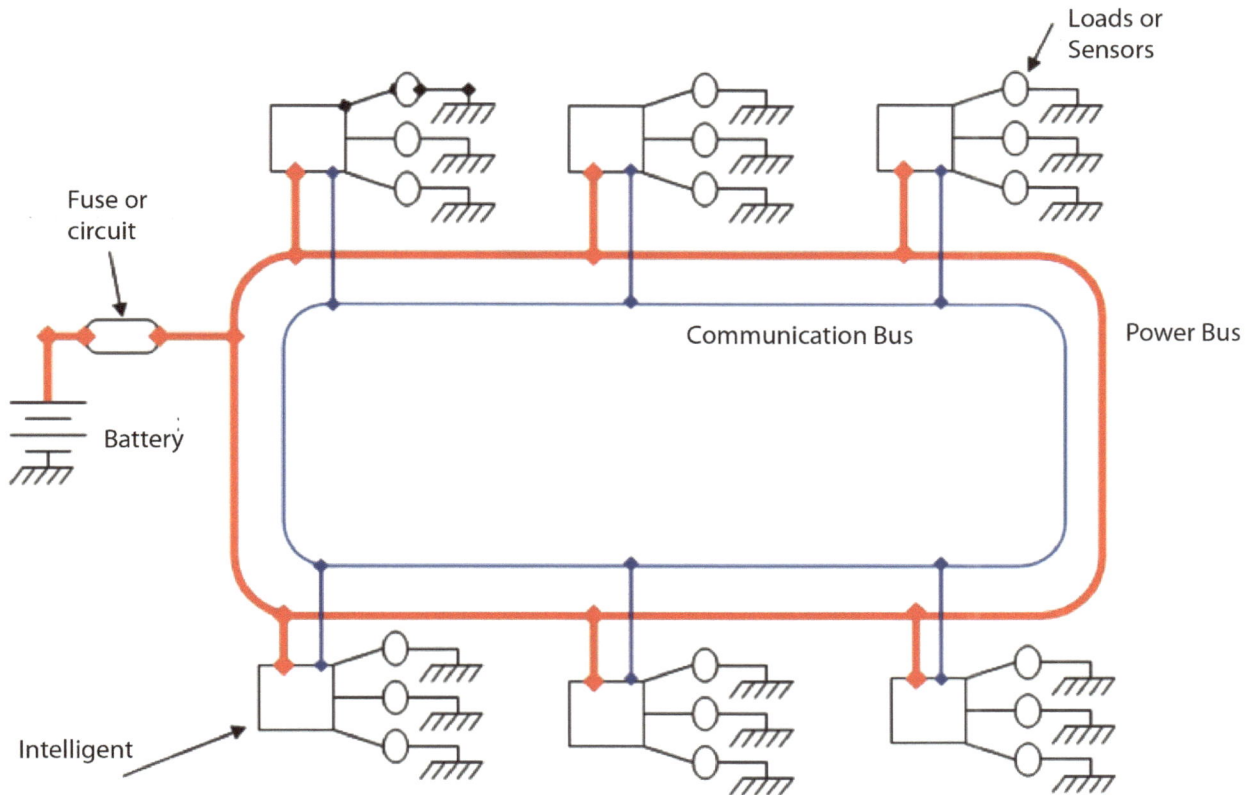

Fig. (1). An automotive multiplex system architecture configuration with power and communication buses, loads, and intelligent nodes for processing communication protocol and load management.

Before delving into this issue, it may be instructive to discuss briefly why an ECU will fail in the first place. Normally the ECU will have sockets to connect the signal bus and the power bus. This can be something comparable to the power and data cables which connect devices in a computer motherboard. The picture of an ECU and a circuit board are shown below. The circuit board in Fig. (**2b**) below is not for the item shown in Fig. (**2a**). These pictures are merely to give an idea of the thing. This could be the engine control unit, which can be very completed. On the other hand there are simpler ones with simple functions.

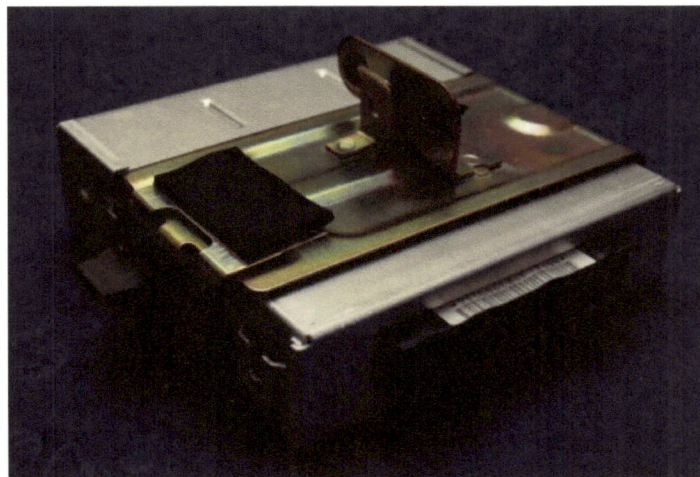

Fig. (2a). ECU picture from outside.

Many times the failure in these devices are due to mechanical reasons like connector getting lose or damaged due to vibration, heat, corrosion, dirt etc., or the solder getting cracked due to thermal fatigue due to repeated heating and cooling. These could sometimes be to items outside of a microprocessor chip and in some cases could be internal problem with the large scale integrated circuits. In addition, there are elements like resistors, capacitors, inductors on the circuit board which could eventually fail after many hours of usage. Some of these causes of failure are preventable by taking good care during usage, but some items are beyond that and will eventually fail regardless.

Fig. (2b). Inside of an ECU.

(a)http://upload.wikimedia.org/wikipedia/commons/6/66/2008-04-17_ECU.jpg   (b)E-book_FT_DBW_   Anwar_r4. docxhttp:// upload.wikimedia.org/wikipedia/ commons/a/a6/ElectronicDieselControlEcuBottomside.jpg

WHAT OTHER FAILURES CAN HAPPEN

In addition to the ECU, other items like cables and connectors outside the ECU could fail as well for similar reasons as before.

All the above are hardware failures. Software failure can happen too, though due to different reasons. Most likely cause of a software failure will happen if the algorithm encounters a situation which was never anticipated before, and which was not considered during design of the algorithm. This can lead to generation of wrong messages or lead to something not activated when asked for, or something suddenly getting activated when not asked for. Both are obviously not very desirable. Or, sometimes one can get into an unending loop, which can be dangerous too.

Hardware failures can, in general, be corrected by replacing the faulty item, or sometimes a whole set of items surrounding the faulty item, together with the faulty item. However, software problem cannot be rectified so easily, but from a user perspective cured tentatively by shutting down the power and resetting the system etc. But sooner or later such an event will need to be referred to the manufacturer to come up with a revised version of the device/software, and is more complex in nature since it cannot be addressed by the dealership immediately.

QUANTITATIVE ANALYSIS OF THE CONSEQUENCE OF FAILURE

As noted earlier, obviously a failure in the ECU node can lead to all loads connected to it to fail. Some failure can be catastrophic or even fatal, like brake or steering system failure. Other failure like an entertainment system malfunction may not be important at all. Some items like window mechanism failure may have security issues etc. So, failure items are not all alike. Hence to analyze such a system a methodology has to be developed. Another thing to be noted here is that it is really not necessary to know microscopic details of all the failure mechanisms to study such systems. Most of the time it is good enough to know an overall (macroscopic) component or subsystem level reliability number which can be quantified by mean time between failure over a period of time or over a given mileage covered. The author discussed these in more detail in his other works (Masrur *et al.*, 1989; Masrur *et al.*, 2003; Masrur *et al.*, 2004; Masrur *et al.*, 2008).

QUANTITATIVE METRIC FOR SYSTEM LOAD AVAILABILITY

As previously noted, the ultimate purpose of the wiring harness (consisting of both the power and the communication buses) is to actuate certain loads, based on the information received from the driver of the vehicle and/or the various sensors, and to coordinate these in a certain desired manner. That is essentially what makes a "drive-by-wire" system. To study the overall system level reliability of such a system, we have to know about all the items from source to load. Hence we can define the following terms.

λ_i = probability that the i-th load/sensor is up (or available) at a particular moment, -- which can range from 0 to 1.

C_i = criticality of the i-th load/sensor, -- a number which indicates how critical it is for this particular load to be up. A range of 1 to 10 is chosen for convenience, where 1 means not at all critical, to 10 meaning absolutely critical.

H_i = number of times on an average the i-th load is invoked or its status updated, during a given span of time (hr, min, sec etc.), -- a number which will depend on the nature of the load. The duration can be chosen to be in units of hour, if it is deemed that H_i may be a small fraction and inconvenient to use.

Based on the above, we can define a figure of merit indicating how detrimental it is for a particular load/sensor (the i-th load/sensor) to be down, by introducing,

n

$$F_i = C_i \, H_i \, (1-\lambda_i) \, \{ \, \prod (\lambda_j) \, \} \tag{1}$$

$$j = 1, \dots n, j \neq i$$

where n is the total number of loads (including sensors) in the system. In equation (1), the term $(1-\lambda_i)$ indicates the probability of the i-th load being down. In this equation we make the assumption that the probability of more than one load being simultaneously down is of second order, and hence much smaller compared to the terms used in equation (1). Based on the above, a cumulative system level figure of demerit can now be defined as:

n

$$F_s = \sum F_i \tag{2}$$

i=1

assuming that the total number of loads = n.

* The above two Sections are based on the author's paper referenced above

ILLUSTRATION OF APPLICATION EXAMPLE

The most important item involved in evaluating the cumulative system figure of merit (or demerit) requires finding the value of λ_i. If we want to evaluate this for the example architecture shown in Fig. **1**, the following items are to be accounted for. This list is just a possible example of items that can lead to failure, and depending on the specific architecture it will vary. But the methodology described here will be valid regardless of the particular case being studied.

Hard Items

1. Battery to battery-cable connector

2. Battery-cable to main-fuse connector

3. Main-fuse to power-bus connector

4. Power-bus cable

5. Power-bus cable to intelligent-node connector (each node consisting of both the power module and also the communication module).

6. Signal-bus (twisted pair etc.)

7. Signal-bus to node connector

8. Node-module

9. Node to load-fuse connector

10. Load-fuse to load connector

11. Load to ground connector

12. Electromechanical relay or solid-state switches connecting the load (can replace the fuse depending on the system).

Soft Items

1. Network message overload and/or error at source end (at message initiating node) causing priority based queuing and leading to delay and/or error in transmission (Masrur, 1989).

2. Failure to win contention with other nodes leading to delay for the message to reach destination, and/or error in message transmission.

It should be noted that the probability of failure of hard items changes with time, starting from infant mortality to deterioration with usage and age. For the soft items, the probability of failure will depend on the message loading and interval used in the system, and is directly related to the quantity $\sum H_i$, for i = 1 to k (number of loads), which was defined earlier. In the above example the numerical values were arbitrarily chosen though based on some reasonable assumptions.

Let us assume for now that there are six nodes in the system with three loads connected to each node. This is the configuration shown in Fig. **1**. The reliability of each item is indicated by the symbol ξ.

For hard items 1 to 7 let us choose: $\xi_1 = 0.99999$ $\xi_2 = 0.99997$ $\xi_3 = .99997$ $\xi_4 = 0.99999$ $\xi_5 = 0.99998$ $\xi_6 = .99999$, $\xi_7 = 0.99999$

For node module itself let us choose $\xi_8 = 0.99995$

For node to fuse connectors let us choose $\xi_9 = 0.99996$ (same for other node to load connectors)

For fuse to load connectors let us choose $\xi_{10} = 0.99996$ (same for other nodes to their respective load connectors)

For load to ground connectors let us choose $\xi_{11} = 0.99996$ (same for other nodes to their respective ground connectors)

For the electromechanical (or solid-state) relays (can replace the fuse where applicable), let us choose $\xi_{12} = 0.99995$

For soft items 1 and 2, let us choose as follows: $\xi_{13} = \xi_{14} = 0.99998$

In the above we just traced the items for one single node. The same will apply to the other loads. For hard items, only the items from 1 to 4 will be common to all nodes, and the rest of the items will be separate for each nodes.

For the 18 loads let us assume the following quantities for C_i and H_i:

$C_1 = 10$ $H_1 = 20$ $C_2 = 10$ $H_2 = 3$ $C_3 = 7$ $H_3 = 10$ $C_4 = 8$ $H_4 = 12$ $C_5 = 1$ $H_5 = 20$ $C_6 = 4$ $H_6 = 20$ $C_7 = 10$ $H_7 = 100$ $C_8 = 3$ $H_8 = 8$ $C_9 = 6$ $H_9 = 14$ $C_{10} = 2$ $H_{10} = 10$ $C_{11} = 1$ $H_{11} = 1$ $C_{12} = 2$ $H_{12} = 20$ $C_{13} = 3$ $H_{13} = 150$ $C_{14} = 9$ $H_{14} = 60$ $C_{15} = 10$ $H_{15} = 2$ $C_{16} = 10$ $H_{16} = 8$ $C_{17} = 5$ $H_{17} = 5$ $C_{18} = 6$ $H_{18} = 40$

For the example case here, the probability that a particular (n-th) load is available is given by the product of all the reliability terms ξ_i for i = 1 to 14.

$$\lambda_n = \prod_{i=1}^{14} \xi_i \qquad (3)$$

In the particular example, if the values of ξ_i are inserted, we get $\lambda_n = 0.99962$, for all n=1 to 18, assuming same component reliabilities in each node. Hence we can easily see that whereas the individual component failure probabilities (ξ_i) were chosen to be between 1 to 5 per 100000, after combining all the components together, we get a failure rate of about 38 per 100000. Although this number is quite small, it might still contribute significantly toward the overall system availability or lack thereof.

In addition to the above, since in general multiple loads are connected to a particular node, there is a reliability issue which leads to the failure of a cluster of loads all together, should any one or more of the linkages leading to the node fail. This can pose a potentially dangerous situation during failure. Thus we also define a group load reliability (or availability) in the following manner. For the n-th node, the group reliability is given by:

$$\lambda_n = \prod_{i=1}^{8} \xi_i \qquad (4)$$

In the example above, this becomes = 0.99983, or 17 failures per 100000, which is about half the individual load failure rate.

It is therefore noticed that the group reliability of the loads connected to node 1 and indicated by equation (4), is somewhat higher than the individual load reliabilities given by equation (3). This is naturally expected, since there are more linkages preceding an individual load than for the group of loads connected to a particular node.

Using these quantities (based on equations (3) and (4)) for all the nodes, we can easily compute that for this 6-node, 18-load system, the Figure of merit (actually demerit to be more exact) according to equation (1) will be as indicated in the following.

Figure of demerit for the system considering individual load failure probabilities is computed by excluding multiple failure probabilities, which is normally of second order, compared to single component failures. So, we compute the probability of failure of load # 1 in node # 1, and assume that all the other loads and nodes are up (available). This leads to, with all the loads taken into account, a cumulative figure of demerit due only to individual load failures to be:

$$18 \quad 17 \quad 6$$
$$F_s = \sum C_k H_k (1-\lambda_k) \ \{ \prod(\lambda_m) \} \ \prod(\lambda_j) \qquad\qquad (5)$$
$$k=1 \quad m=2 \quad j=1$$

where λ_k for k-th load is evaluated using equation (3), and λ_j is evaluated by equation (4). λ_m is evaluated equation (3), but slightly modified, so that the reliability of elements already included in λ_j are excluded here (in other words, only components 9 to 14 are now included during this evaluation, rather than 1 to 14, as in equation (3)). Inserting all the necessary values, we get from equation (5),

$$F_s = 0.7677 \qquad\qquad (6)$$

In equation (5), the value of j runs from 1 to 6, for the six separate clusters of loads. Since this system also has the possibility of group failure as indicated earlier, we have to account for that as well. This is computed by using equation (1), except that we now use equation (4) instead of equation (3) to compute the λ_j and this λ_j (for the j-th cluster) will correspond to each cluster of loads corresponding to each node, there being a total of six such clusters in the example case. During computation of the group failure, we assume that only one cluster can fail at a time. Thus the equation for computation of cumulative failure probability, assuming all the clusters can fail one at a time is given by:

$$18 \quad 6$$
$$F_s = \sum C_k H_k \ [(1-\lambda_j)]_{j=1} \prod(\lambda_j) \qquad\qquad (7)$$
$$k=1 \quad j=2$$

Here, first we compute the probability of cluster # 1 being down and others up, but since the result is symmetric, we can extend it to others to compute the cumulative figure of demerit. Thus, in our example case this group failure figure of demerit will become = 0.3448. Hence adding this with equation (6) we get the next level of figure of demerit for the system to be = 1.1125. Finally there is the probability of a total system failure due to the hard linkages 1 to 6. Following the same line of thought, it leads to an additional figure of demerit = 0.2233. Hence if we add this number to the previous ones, the final cumulative system figure of demerit will be:

$$F_s = 1.3358 \qquad\qquad (8)$$

which is a combination of items due to figure of demerit from individual load failures (0.7677), figure of demerit from group load failures (0.3448), and the figure of demerit from total system failure (0.2233). It can be immediately observed that the contribution to the total system figure of demerit comes mostly from the individual load failures, next from a group of loads failing together, and least from the total system failure. This is naturally expected. To best capture the phenomena of total system, group and individual failures, it is rational to add these numbers as has been done to get the equation (8), since all of these contribute towards the demerit of the overall system. This number can therefore be used as a metric for the purpose of evaluating the quality of a system, and the larger the number, the worse the system availability.

WHEN TRYING TO IMPROVE, IT IS IMPORTANT TO HAVE A HOLISTIC VIEWPOINT OF THE SYSTEM

In the previous example (though used assumed items involved in the by-wire system, and assumed numerical values for reliability numbers) it is clear that several things contribute to system failure. Obviously then, to improve the system reliability and robustness, all of those items have to be "improved". The followings could be done to achieve that goal.

Quality enhancement at component level - by increasing this we are basically increasing the reliability number ξ_i noted previously for various components. However, to increase these numbers like 0.99995, it becomes extremely costly as one tries to increase the value at higher decimal points. This means, in practical terms, that to increase the value of ξ_i from 0.999 to 0.9999 is relatively costlier than trying to increase the number from 0.99 to 0.999. So, at one point component quality enhancement may be impractical.

Introducing redundancy – a very important method to increase the overall system level reliability is to have additional components with parallel functionality. Even though the components themselves may have same reliability, the overall probability of failure is substantially reduced by redundancy. For example, if a component has $\xi_I = 0.99$, then by having two such items in parallel the overall reliability becomes {1- (1-0.99)(1-0.99)} = 1-(.01)(.01) = 0.9999, under the assumption that overall failure requires both of the components to fail simultaneously. To achieve such a degree of reliability enhancement by merely using one component and improving its quality would be very difficult. This is a case of system architecture modification. Redundancy is a very important mechanism, which is abundantly noticed in nature e.g. biological system like human brain.

Software reliability – software reliability can be compromised primarily due to human error during design by not being able to account for every foreseeable circumstances that an algorithm might be subjected to. This can, to some extent, be mitigated by introducing dual paths for an algorithm where the algorithms in these dual paths can be developed by different software groups. Through some voting scheme, if two or more software paths are in agreement, the software function may be considered correct.

The above methods involving hard and software redundancy are indicated through the block/flow diagram below between the system input and output as shown in Fig. (**3**).

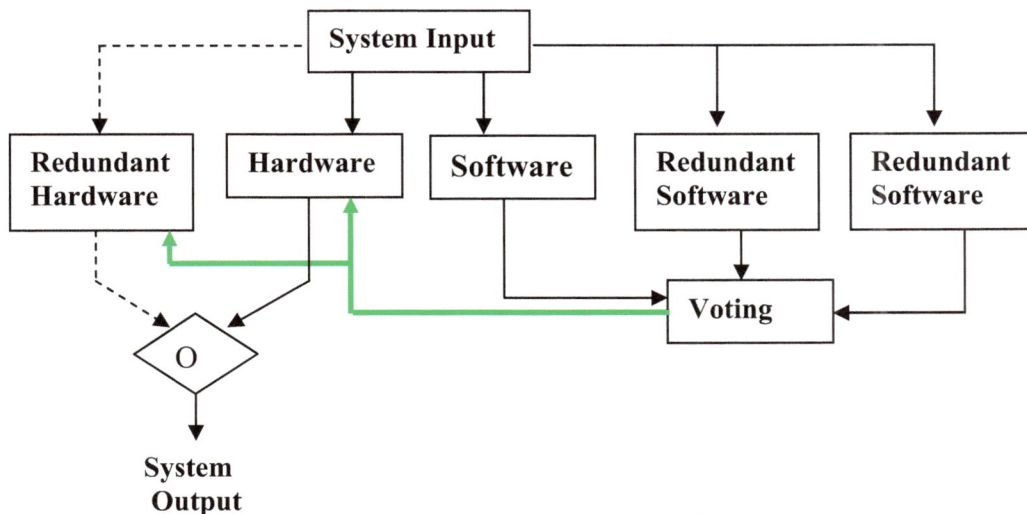

Fig. (3). System level hardware and software redundancy.

Architectural reconfiguration for reliability enhancement – In a by-wire system, especially for safety critical items, it is extremely beneficial to have redundancy of the type indicated in the above diagram, which can have different variants. If a system encounters a fault, the above redundancy can make it "fail safe" or "fail silent", to the extent of a single failure. In the system above, if there is a software failure in one of the 3 software blocks, it can still continue to function. After that there won't be any way to know through voting scheme which software is correct. Since software does not pose space issue in terms packaging or physical cost, it is easier to introducing redundancy. For hardware it is difficult to do so. Hence in case of hardware one cannot normally afford to have more than one extra.

Normally if a malfunction is detected through any diagnostic scheme, or if a future risk is identified through prognostics, it will become necessary to use the duplicate or redundant hardware and disconnect or isolate the faulty one,

Multi-system separate network architecture with gateways when beneficial – A by-wire system, depending on its situation can sometimes be partitioned into multiple ones with different network protocols to carry information. For example the unimportant or less safety critical loads can be controlled by a CAN network and the safety critical ones may be kept under a more deterministic network like a TTP (time triggered protocol). Depending on the needs, there can be some kind of gateway between the two networks to exchange information between the two. This is shown in the diagram illustrated in Fig. (**4**) which is a revised version of the previous diagram, and indicates a partitioning of the loads into two different network buses, one catering more safety critical loads than the other.

Failure due to non-network reason - As is obvious, in our analysis of reliability above, we have taken into account the hardware components, software, and communication network into consideration. Through this process it has been emphasized that overall system reliability is dependent on the complete and successful working of everything in unison, and not just the communication network in a DBW. It is the intent of this section to drive this point.

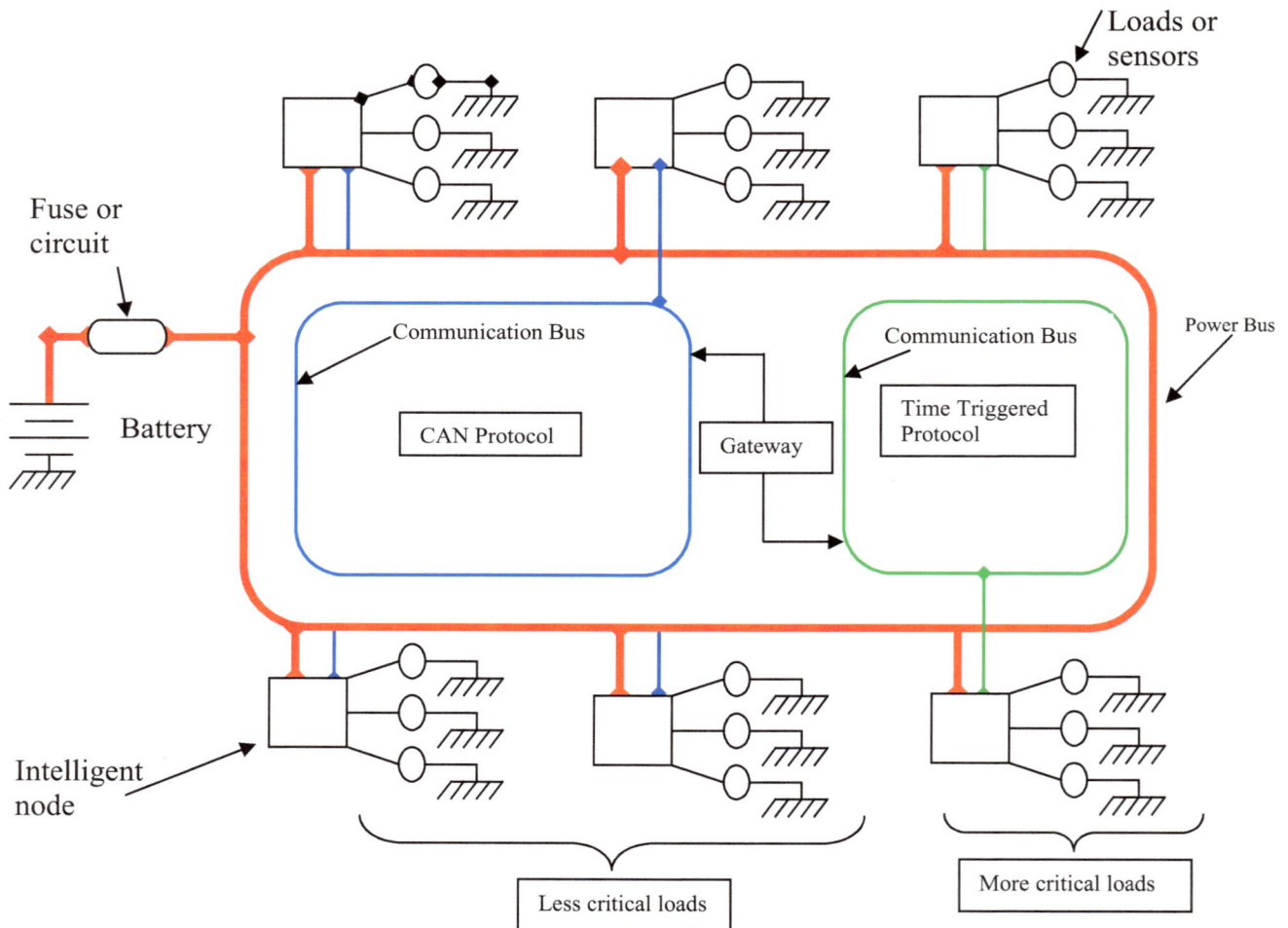

Fig. (4). Multi-protocol communication with gateway.

CONCLUDING NOTES

The main aim of this section has been to take a complete system level perspective towards reliability of drive-by-wire systems. The intent was to create a generic framework applicable to study any DBW. The author has extended the above ideas to HEV's (hybrid electric vehicular systems) and provided a quantitative metric to compare between different HEV's against regular IC engine vehicles (Masrur, 2008). From the discussion in this section, it is seen that the overall system reliability (for DBW and also non-DBW) comes initially through the quality of components used, but at some point a limitation arrives in quality enhancement due to cost and other fundamental limitations. At that point it becomes important to use redundancy in terms of both hard and software. The treatment of analysis for both DBW and non-DBW can be done by using the general methodology discussed in this section. It should also be emphasized here that it is not possible to completely eliminate system failure. However by using good diagnostic and prognostic methods (which is not within the scope of this section) it is in many cases possible to know about the health of a system before a failure occurs, and take preemptive steps to avoid those failures, which is of great importance, particularly for safety critical systems.

REFERENCES

Arora, A., Ramteke, P., & Mahmud, S. (2004). Fault Tolerant Time Triggered Protocol For Drive-by-Wire Systems. Proceedings of the 4th Annual Intelligent Vehicles Systems Symposium of National Defense Industries Association (NDIA), National Automotive Center and Vetronics Technology, Traverse City, Michigan.

Isermann, R., Schwarz, R., & Stolzl, S. (2002, October). Fault-tolerant Drive-by-Wire Systems. IEEE Control Systems Magazine, 64-81.

Masrur, M.A. (1989). Digital simulation of an automotive multiplexing wiring system. *IEEE Trans. on Veh. Tech.*, 38(3), 140-147.

Masrur, M.A. (2008). Penalty for Fuel Economy – System Level Perspectives on the Reliability of Hybrid Electric Vehicles During Normal and Graceful Degradation Operation, *IEEE Systems Journal*, 2(4), 476-483.

Masrur, M. A., Garg, V. K., Shen, J., & Richardson, P. (2003). Comparison of System Availability in an Electric Vehicle with Multiplexed and Non-Multiplexed Wiring Harness. IEEE Vehicular Tech. Conf. Proceedings.

Masrur, M.A., Shen, Z.J., & Richardson, P. (2004). Issues on Load Availability and Reliability in Vehicular Multiplexed and Non-Multiplexed Wiring Harness Systems. Society of Automotive Engineers (SAE) Transactions, Journal of Commercial Vehicles, 2003-01-1096, 31-39.

Seidel, F. (2009). X-by-Wire. Hauptseminar Transportation Systems, Chemintz Univ. of Tech., Feb, 1-9.

Wikimedia Page. Retrieved on 06th June, 2011, http://upload.wikimedia.org/wikipedia/commons/6/66/2008-04-17_ECU.jpg

Wikimedia Page. Retrieved on 06th June, 2011, E-book_FT_DBW_Anwar_r4.docxhttp:// upload.wikimedia.org/ wikipedia/ commons/a/a6/ ElectronicDieselControlEcuBottomside.jpg.

CHAPTER 3

Dependability and Functional Safety

Giuseppe Buja[1*] and Roberto Menis[2]

[1]*Department of Electrical Engineering, University of Padova, Via Gradenigo 6/a, 35131 Padova, Italy and* [2]*Department of Electrotechnics, Electronics and Computer Science, University of Trieste, Via Valerio 10, 34127 Trieste, Italy*

Abstract: The chapter deals with the dependability and the functional safety of a system by illustrating the key points of the theoretical corpuses formulated on the two subjects: the dependability theory and the functional safety standards. Dependability is concerned with the ability of a system to deliver the intended service, including the ability to cope with a fault. Functional safety is concerned with the safety-critical systems and focuses on the characteristics of the extra systems added to a system with the purpose of making its operation safe. The chapter starts by providing the definitions of system and service. Then it passes to the illustration of the key concepts of the dependability theory, which are the threats, the attributes, and the techniques used to enforce the dependability. Particular attention is given to the fault-tolerance techniques and the architectures of the fault-tolerant systems. Afterwards, the chapter presents the key issues of the functional safety standards, which are the analyses of hazard and risk of a safety-critical system, and the safety requirements for the extra systems. At last, a case of study is examined from the standpoints of both the dependability and the functional safety.

Keywords: Functional Safety Standards, Dependability Theory, Fault Tolerant Systems, Safety Critical Systems, Steer By Wire, Fault Tolerant Architectures, Mean Time to Fail (MTTF), Mean Time to Repair (MTTR), Probability of Failure on Demand (PFD).

INTRODUCTION

In the modern society, an increasing number of artificial systems -hereafter briefly termed systems- is utilized to support the human activities. As a result, these activities depend more and more on the services delivered by the systems and this is perceived especially when the intended services are not delivered correctly or, simply, are not done. Furthermore, the complexity of the systems is getting higher and the chance that a fault prevents the systems from delivering the correct services is becoming greater than ever. This event is particularly crucial for safety-critical systems, i.e. for systems whose incorrect service or, at worst, outage can damage people, things or the environment. The two subjects so far addressed, namely the ability of a system to deliver the intended service and the ability of a system not to produce damages under a fault are referred to as dependability and safety of a system (Storey, 1996; Avizienis *et al.*, 2001; Avizienis *et al.*, 2004; Buja & Menis, 2007; and Al-Kuwaiti *et al.*, 2009).

Dependability and safety are subjects that have aroused a large interest in the last decades, mainly stimulated by the broad usage of electronic devices in the control stage of a variety of systems. The underlying reason is that the electronic devices are thought to be less robust in withstanding adverse situations, such as the physical stresses, than the mechanical or hydraulic devices that they often substitute for. This explains the specific attention that the developers have paid over time to the electronic equipment embedded in a system. For instance, just in 1967 A. Avizinies formulated the fault-tolerant concept stating "we say that a system is fault-tolerant if its programs can be

***Address correspondence to Giuseppe Buja:** Department of Electrical Engineering, University of Padova, Via Gradenigo 6/a, 35131 Padova, Italy; E-mail: giuseppe.buja@unipd.it*

properly executed despite the occurrence of logic faults" (Avizienis, 1997). The concept was tailored to the computer-based systems, where the services are the execution of programs, and was formulated at the time when the computers started to be introduced in the control of nuclear plants. Since then the theoretical knowledge on the matter has advanced, arriving at the formulation of theoretical corpuses on the dependability and on a safety-correlated subject, namely the functional safety. The two corpuses are the dependability theory and the functional safety standards. In parallel with such a formulation, the computer-based systems have entered virtually everywhere in the human activities like in civil appliances, industrial manufacturing and transportation, just to mention a few.

The dependability theory is focused on the characteristics of the service delivered by a system, including those relevant to safety, while the functional safety standards are focused on the characteristics of the means added to a system to assure its safety, where the means are constituted by hardware and/or software components. The two corpuses are strictly correlated when they deal with a fault and with its effects on the safety as well as when they have recourse to additional means to cope with a fault. The additional means are commonly termed as extra system since they are not necessary to deliver the service.

In this chapter the dependability theory and the functional safety standards are explicated with the purpose of providing a comprehensive treatment for the analysis of a system from the points of view of the dependability and the functional safety. In particular, it is shown how one corpus supplements the other one. At first the dependability theory is introduced and the relevant concepts of threatens, attributes and enforcing techniques are presented. With regard to the safety, the dependability theory limits itself to the enunciation of principles without paying attention to their application to the systems. This gap is filled up by the functional safety standards, which are introduced in the second part of the chapter together with the concepts of risk, safety function and safety integrity level. Finally, as a case of study, the subjects of dependability and functional safety are exemplified with reference to the power converter supplying the actuator of a steer-by-wire system.

SYSTEMS AND SERVICES

A system is a set of components grouped together into a single entity with the purpose of delivering a service, where the service is the outcome of the functions carried out by the components. Requirements for the service that the system must deliver are usually collected in a document. For safety-critical systems, there is a further set of requirements related to safety; they are also collected in a document, which explains what the system must do or not do under a fault to behave safely.

A composite system can be thought to be made up of a number of simpler systems, denoted with subsystems, which in turn are made up of other subsystems and so on: the recursive process stops when a subsystem is considered as an elementary one.

Any system interacts with other entities. Some of them, commonly termed feeders, solicit the system whilst other ones, commonly termed users, utilize the service delivered by the system. Users and feeders can be both systems and human beings. Often a system consists of a number of subsystems that, as a whole, deliver a service to a human user. In this circumstance, the subsystems exchange with each other intermediate services and the human user receives the end service of the system.

The behavior of a system is represented by its state as a function of time, where the state is a set of modes of operation and values of physical quantities. The state can be separated into two sets: the external state, which is the portion of state released by a system to the user, and the internal state, which is the portion of state transparent to the user. As a matter of fact, the service delivered by a system is correct when the external state meets the appointed requirements, which are expressed in a qualitative form for the modes of operation and in a quantitative form for the values of physical quantities.

The functions underlying a service can be executed by a single system or can be shared among individual subsystems connected by means of a communication network. The first solution gives rise to a function-concentrated system, the second one to a function-distributed system (or, more concisely, distributed system). In a distributed system, a subsystem exchanging data with another subsystem through the network is termed node. Driven by the introduction of effective communication networks, the distributed systems have gained momentum because they facilitate the implementation of complex and/or large systems.

DEPENDABILITY THEORY

Key concepts of the dependability theory are threatens, attributes and enforcing techniques (Storey, 1996; Avizienis *et al.*, 2001; Avizienis *et al.*, 2004; and Al-Kuwaiti *et al.*, 2009).

A) Threatens

From the dependability perspective, a system can be represented by the three-layer model of Table **1** where each layer is identified by the activity done. The threatens of the dependability are of three kinds, depending on the layer where they occur, and are termed faults, errors and failures. A fault is a deviation of the operation of a component from the expected one; when it is caused by physical phenomena like mechanical and/or electrical stresses, wear, aging and heating of the component, the fault is termed internal. Other threatens assimilated to a fault are i) the failure of a feeder and the mistake of a human being that solicits the system; this type of fault is termed external, and ii) a flaw in the development stage of the system that can be, for instance, an oversight in the component sizing or a bug in the software routines. A conceivable, context-dependant classification of the faults is reported in Table **2**. An error is a deviation of the internal state of a system from its true value. Errors may occur only in the systems that process information items like data or signals. This is, for instance, the case of a control system whether based on electronic technologies or not. A failure is a deviation of the external state of a system from the appointed one and produces a system outage. A conceivable, context-dependant classification of the failures is listed in Table **3**.

Table 1. Dependability-Related Model of a System

LAYER	THREATEN
OPERATIONAL (service delivering)	FAILURE
PROCESSING (information processing, if any)	ERROR
PHYSICAL (component operation)	FAULT

Table 2. Fault Classification

CONTEXT	FAULT TYPE
Nature	Hardware fault
	Software fault
Persistence	Permanent fault (a fault continuously or regularly recurring in the time)
	Transient fault (a fault bounded in time and not recurring)
	Intermittent fault (a transient fault whose activation is not systematically reproducible)
Activity	Dormant fault (a fault not causing an error)
	Active fault (a fault causing an error)
	Latent fault (an active fault not yet manifested or detected)

Table 3. Failure Classification

CONTEXT	FAILURE TYPE
Domain	Content (or value) failure
	Time failure
Propagation	Halt failure (a failure stopping the system service)
	Erratic failure (a failure not stopping the service and then possibly propagating or generating a fault in the fed systems)
Detectability	Signaled failure (a failure originated from an error detected and revealed by a warning signal)
	Unsignaled failure (a failure originated from a non-detected error and then not revealed by a warning signal)
Perception (appropriate when the system has two or more users)	Consistent failure (a failure identically perceived by the users)
	Inconsistent (or Byzantine) failure (a failure differently perceived by the users)
Consequences (or severity)	Benign (or harmless) failure
	Negligible, marginal, critical failure
	Catastrophic failure

Fig. (**1**) illustrates the two basic mechanisms of failure: generation and propagation. The rule holds that a failure is generated by an error and an error is generated by a fault. Therefore a fault is the triggering source of a failure according to a cascade generation mechanism. When the system has no processing layer, a failure is directly generated by a fault. Not all the faults generate an error but sometimes they stay confined in the respective layer. The same takes place for the errors and the failures. In this sense, an error is the emergence of a fault in the processing layer, a failure is the emergence of an error in the operational layer, and an incorrect service is the emergence of a failure to the users. Faults, errors and failures can be also caused by other faults, errors and failures through a propagation mechanism located within the same layer.

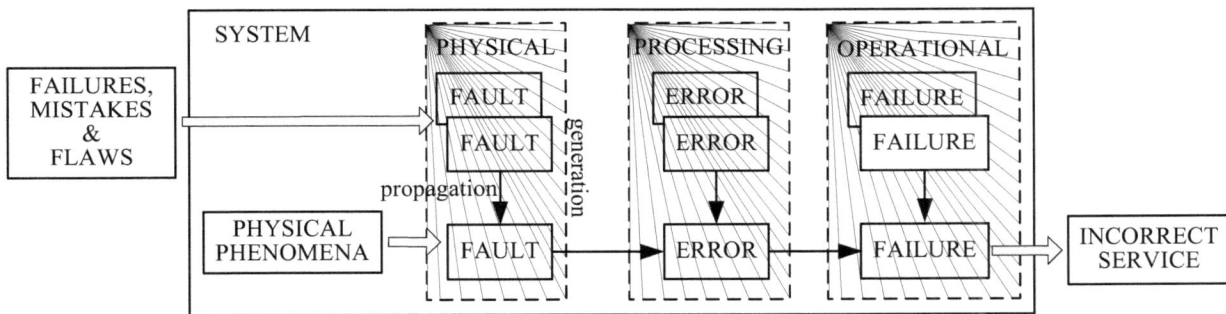

Fig. (1). Failure mechanisms.

B) Attributes

The attributes are quantities that measure the dependability of a system. The main attributes are reliability, availability, maintainability and safety. The first three attributes are expressed by probabilistic figures, defined as follows:

- **Reliability** is the probability that a system carries out the correct service at the time t>0 for a given set of operating conditions, provided that the correct service is done at the time t=0. High reliability means long time intervals before the system fails. The expected time for a system to fail is the Mean Time To Fail (MTTF). It is a statistical value and, therefore, the length of the observation interval needed to determine the MTTF should be infinite.

- **Maintainability** is the probability that a system delivers the correct service at the time t>0, provided that the correct service is not done at the time t=0 and a repair process is in progress. High maintainability means short downtime for the system. The expected time for a system to be repaired is the Mean Time To Repair (MTTR). It is also a statistical value.

- **Availability** is the probability that a system delivers the correct service at the time t>0, without specifying whether the service is correct or not at the time t=0. High availability means long uptime of the system. As a function of the MTTF and the MTTR, the availability is expressed as,

$$A = \frac{MTTF}{MTTF + MTTR} \tag{1}$$

By the dependability theory, safety is a property that a system either has or does not have. Safety is important for safety-critical applications and is the ability of a system to show a safe behavior in the presence of a fault that yields a non-acceptable failure.

C) Enforcing Techniques

The enforcing techniques are intended to improve the dependability of a system, safety included; they act by preventing, tolerating, removing and forecasting a fault. In Table **4** the techniques and their actions are detailed. The fault-tolerance techniques are the most popular ones since, differently from the other techniques, they act while the system is operating and makes it tolerant to a fault (Nelson, 1990).

The requirements in terms of fault-tolerance are classified into three levels, commonly known as fail-operational, fail-safe and fail-silent levels, where the generic term "fail" embraces faults and errors. By the fail-operation level, the system continues to deliver the correct service in spite of a fail, thus enhancing both the reliability and the safety of the system. The fail-operational level is needed when an uninterrupted service is demanded. A particular case of fail-operational level is the fail-degraded operation that refers to the case of a system that delivers a correct but downsized service. By the fail-safe level, the system responds to a fault by reaching a safe state; the fail-safe level is equivalent to a benign failure, where the difference is that here it is obtained by using fault-tolerant techniques. By the fail-silent level, the system goes off after a fault; this fault-tolerance level is of interest when the faulty system either delivers a non-critical service or produces a fault in a fed system.

Table 4. Dependability Enforcing Techniques

Enforcing techniques	Action
Fault-prevention	Aimed at avoiding the occurrence of a fault. Applied during the design, development and test stages of the system.
Fault-tolerance	Aimed at coping with a fault. Applied during the system operation and implemented by redundancy.
Fault-removal	Aimed at finding and eradicating a fault. Applied after the system setup to verify the fulfillment of the dependability requirements. Able to start diagnosis and corrective procedures.
Fault-forecasting	Aimed at evaluating the dependability performance. Based on measurement or computation of the attributes.

FAULT-TOLERANCE TECHNIQUES

The fault-tolerance techniques exploit the redundancy principle, i.e. the insertion into the system (or into a subsystem) of an extra system that contrasts the faults. An extra system that exactly reproduces a part of the system is termed replica. When using a replica, two basic schemes can be arranged: passive and active. By a passive replication, the system delivers the service and one (or more) replica serves as a backup that starts working only when a fault occurs in the system. By the active replication, one (or more) replica works in parallel with the system and, under a fault in the system, the replica replaces the system. The active replication scheme is fast but requires some determinism of the replica in the time and space domains (Poledna, 1996). Instead, the passive replication scheme does not need any determinism but the replica must be able to reproduce the state of the system at the time of the fault occurrence.

There are essentially two strategies of tolerating a fault, namely the system reconfiguration and the fault masking. The techniques based on the system reconfiguration work in three steps that are i) fault detection, by which the presence of a fault is revealed, ii) fault location, by which the source of the fault is found, and iii) system recovery, by which the detected fault is eradicated and the service is restored through the system reconfiguration. The techniques based on the fault masking ride through a fault by using a replication scheme and do not take care of detecting the fault and recovering from it.

FAULT-TOLERANT ARCHITECTURES

The basic entity of a dependable system is the Fault Tolerant Unit (FTU). FTU is a system that provides one of the three levels of fault tolerance described in Section III. FTU utilizes either the system reconfiguration or the fault masking strategy or both (Buja & Menis, 2007; Nelson, 1990).

A typical fault masking FTU has the Triple Modular Redundancy (TMR) architecture of Fig. (**2**), with three replicated modules and one Voter in cascade that votes the external states of the modules. The architecture exhibits a fail-operational level of tolerance to a fault in one of three modules but has the drawback that the Voter is a single point of failure (i.e. it is a subsystem that, if failing, causes a non correct service of the FTU). The architecture of Fig. (**3**) overcomes this drawback by using three replicated Voters and by allocating them at the input of the scheme. Fault of any subsystem, whether a module or a Voter, is masked by the Voters in the subsequent FTU.

In some applications such as in the automotive field where weight and costs must be kept as much low as possible, the TMR architecture is too heavy and expensive. A more compact architecture consists in arranging an FTU with a Dual Modular Redundancy (DMR) scheme having fault-tolerance capabilities. A simple scheme implementing this architecture is shown in Fig. (**4**). The Fault Detector & Locator performs the fault detection and location by searching for a possible fault in one of the two modules (Isermann & Ballé, 1997). This is done by checking directly the devices within the module or by detecting the effects of a fault. The Selector performs the system recovery by selecting the output of the correctly operating module. The scheme still exhibits a fail-operational level of tolerance to a fault in a module but has again the drawback that both the Fault Detector and Selector are single points of failure.

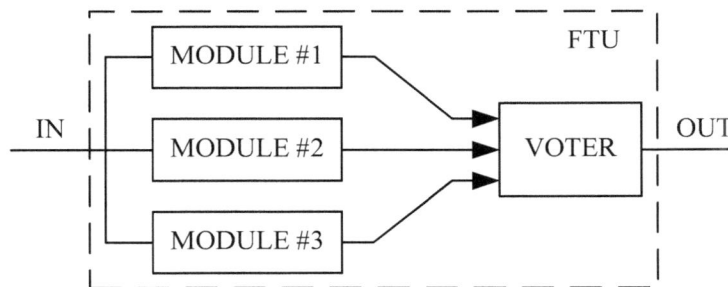

Fig. (2). Typical TMR-based FTU architecture.

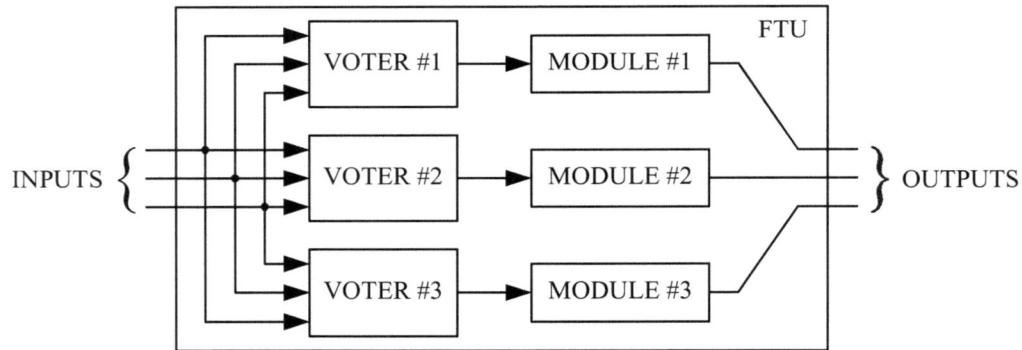

Fig. (3). TMR-based FTU architecture without single point of failure.

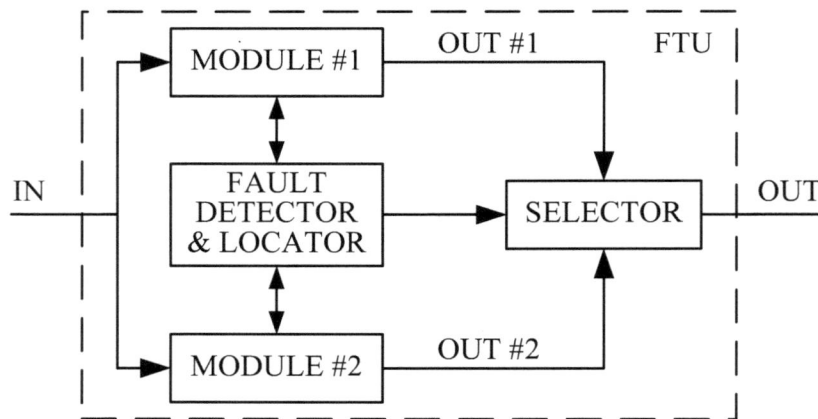

Fig. (4). DMR-based FTU architecture with fault-tolerance capabilities.

A more powerful fail-operational FTU is obtained by putting together two fail-silent FTUs as shown in Fig. (**5**). Each fail-silent FTU is constituted by a module endowed with fault detecting capabilities and a switch that is enabled by the Fault Detector through the no-error signal. The Self Selector recognizes when one FTU goes silent and, under this event, forwards the output of the alive FTU.

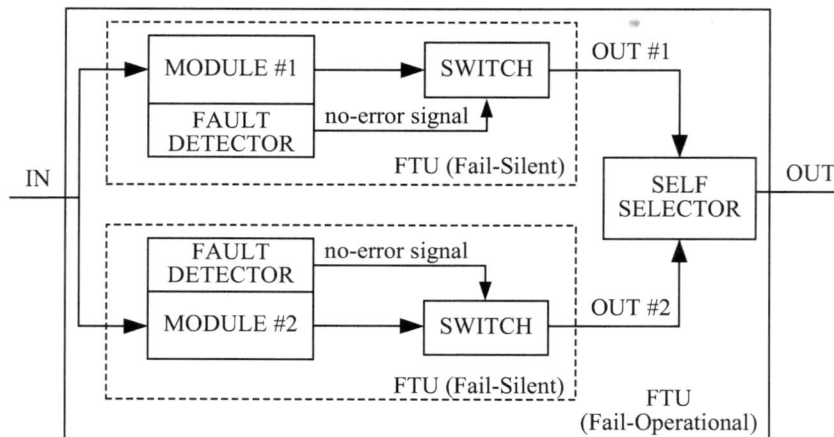

Fig. (5). Fail-operational FTU obtained with two fail-silent FTUs.

FUNCTIONAL SAFETY STANDARDS

The functional safety is concerned with the part of the overall safety of a system that depends on the correct service of the extra systems introduced to make the system safe (IEC TC 65, 1998; Purewal & Waldron, 2004). Before entering into the functional safety standards, it is useful to point out that the standards designate the system with Equipment Under Control (EUC), the extra system in charge of the safety with safety-related system, and faults and failures, when compromising the safety of the EUC, with hazards and accidents, respectively. This terminology is adopted hereafter. By definition, a hazard is a potential source of harmful consequences for people, things or the environment whilst an accident is an event unacceptable for the safety and is consequential to a non-deliberated hazard.

The functional safety standards encompass two key issues: i) hazard and risk analyses, and ii) safety requirements.

A) Hazard and Risk Analyses

The hazard analysis determines the hazards and the consequent accidents occurring in an EUC during its operation. There are several methods to analyze the hazards; the main methods are listed in Table **5**. Some of them like FMEA and FTA are also employed in the dependability framework to investigate the failure mechanisms.

Table 5. Hazard Analysis Methods

METHOD	DESCRIPTION
Preliminary Hazard Analysis (PHA)	Qualitative method that analyzes the sequence of events that could transform a hazard into an accident.
Hazard & Operability (HAZOP)	Qualitative method that identifies the deviations of the EUC state from the normal operating conditions.
Failure Mode and Effect Analysis (FMEA)	Qualitative bottom-up method that starts from a hazard to determine the possible accidents produced by it. Physically oriented method that can be quantified.
Fault Tree Analysis (FTA)	Qualitative top-down method that traces back from an accident to its causes (hazards). It is used for modeling specific accidents and can be quantified.

The risk is a measure of the dangerousness of an accident. It is given in either qualitative or quantitative form. In a qualitative form it is expressed by a linguistic variable obtained by sorting the probability of occurrence (or frequency) and the severity of an accident in a number of classes, and by naming each class with an appropriate term. As an example, the frequency classes are typically named frequent, probable, occasional, remote, improbable and singular whilst the severity classes are typically named catastrophic, critical, marginal and negligible. Frequency and severity of an accident are termed risk probability and risk severity, respectively. Numerous techniques have been devised to express the two variables in a quantitative form (Birolini, 2003). The most popular technique to calculate the risk probability is based on the Markov models whilst the risk severity is commonly represented by a natural number in an arbitrary scale, with the severity that increases by moving from the lowest value to the highest one of the scale. These two variables play an important role in the risk analysis as their product gives the risk. Depending on the consequences allowed for an accident, the risk can be mapped into three regions as reported in Table **6**.

Table 6. Risk Regions

RISK REGION	DESCRIPTION
Intolerable	Risk so great that is absolutely intolerable and must be reduced.
Totally acceptable	Risk so small that is insignificant and no reduction is needed.
Tolerable or Risk As Low As Reasonably Practicable (ALARP)	Risk falls between the two previous regions. The risk is tolerable when it is so low that the cost of a (further) reduction of it is not balanced by the resulting benefits.

B) Safety Requirements

If the risk associated to a hazard is not tolerable, safety functions suitable to reduce the risk must be arranged. They are specified by the following requirements: a) safety function requirements, and b) safety integrity requirements.

The safety function requirements are deduced from the hazard analysis and describe the safety functions to be set up and what they are meant to do. The safety functions are aimed at making the EUC operation safe under the occurrence of a hazard and are implemented in the so-called safety-related systems. These systems can be realized by means of different technologies (mechanical, hydraulic, pneumatic, and so on); the systems relying on the electric/electronic/programmable electronic (E/E/PE) technology are standardized by IEC 61508. A safety function can be either implemented in a system different from the EUC or embedded in a part of the EUC; in the latter case, that part becomes a safety-related system for the safety function implemented. As an example, the control unit of an electric drive becomes a safety-related system if it implements a safety function that impedes the runaway of the drive speed.

The safety integrity is defined as the probability that a safety function is correctly performed by a safety-related system. The safety integrity requirements specify the demands of safety integrity for the safety functions. Formulation of these requirements is necessary to account for situations where the safety-related system fails. Under this occurrence, the implemented safety function is not executed, leaving the EUC not more protected against the hazards faced up by the safety function and the consequent accidents (Carey & Purewal, 2001; Burcsuk, 2007).

The safety integrity requirements are determined from the outcome of the risk analysis and are expressed in terms of Probability of Failure on Demand (PFD) of the safety-related systems. The extreme hypotheses are as follows: i) the PFD of a safety-related system is 0; then the safety function is performed, an hazard covered by the safety function does not generate an accident, and the risk is reduced to zero, and ii) the PFD is 1; then the safety function is not performed, an hazard can generate an accident, and no risk reduction is obtained.

PFD may be found by way of either quantitative or qualitative methods. The quantitative methods rely on the implementation of the following procedure for each hazard:

- Designation of the tolerable risk,

- Calculation of the risk associated to the hazard for the EUC without any safety-related system,

- Comparison of the calculated risk with the tolerable one; if the calculated risk is greater than the tolerable one, a risk reduction is necessary,

- Determination of the magnitude of risk reduction by dividing the tolerable risk by the calculated one; the resulting value gives the PFD of the safety-related system.

The qualitative methods split the PFD range into four levels, termed Safety Integrity Levels (SILs), and determine the PFD in an indirect way by assigning the SIL. Table **7** gives the correspondence between PFDs and SILs (Isermann & Ballé, 1997). The SIL is assigned by help of tools such as tables, graphs or matrices calibrated to the specific EUC and built up on the basis of the tolerable risk for an accident. The techniques used to assess the risk and the relevant SIL are listed in Tables **8, 9** and **10** are examples of SIL assignment by the Consequence Only and Risk Matrix techniques.

Table 7. Correspondence between SILs and PFDs

PFD	SIL
$\geq 10^{-9}$ to $< 10^{-8}$	4
$\geq 10^{-8}$ to $< 10^{-7}$	3
$\geq 10^{-7}$ to $< 10^{-6}$	2
$\geq 10^{-6}$ to $< 10^{-5}$	1

Table 8. Techniques Used to Assess the Risk and the Relevant SIL

TECHNIQUE	DESCRIPTION
Consequence Only	Simplest and most conservative technique. It uses only an estimation of the risk severity whilst does not consider the risk probability. Hazards resulting in an equal risk severity have the same SIL irrespectively of the risk probability.
Modified HAZOP	Extension of the HAZOP method. It assigns the SIL by means of a qualitative risk evaluation and a consequent selection of the SIL that is perceived appropriate by a risk estimation team. It is a very subjective technique.
Risk Matrix	One the most used technique. It uses a two-dimension matrix; each element of the matrix is the SIL correlated to the risk severity (matrix row) and to the risk probability (matrix column).
Risk Graph	Alternative to the risk matrix. It evaluates the risk -and from it the SIL- by means of other risk factors, in addition to the risk probability and the risk severity (for example, the degree of exposure to the accident or the possibility for an user to escape from the accident).

Table 9. SIL Assignment by the Consequence Only Technique

HAZARD CONSEQUENCE	SIL
Potential for very many fatalities	4
Potential for multiple fatalities	3
Potential for single fatality or serious injuries	2
Potential for minor injuries	1

Table 10. SIL Assignment by the Risk Matrix Technique (NR Stands for No risk Reduction)

Risk severity	Catastrophic	3	3	4
	Critical	2	3	3
	Marginal	1	2	3
	Negligible	NR	1	2
		Low	Moderate	High
		Risk probability		

CASE OF STUDY

As a case of study, the actuation module of the Steer-by-Wire (SbW) system of a vehicle is considered (Buja *et al.*, 2004). The SbW system is an all-electric system composed by transducers and motors for entering, sensing and actuating the driving commands, and by a communication network for their transmission. As for the traditional equipment, the main requirements for a SbW system are gradualness, accuracy and dependability of the driving maneuvers. The attributes of the dependability, especially the safety, are undoubtedly very severe for the SbW systems. Indeed, it is evident that a wrong operation or, in the worst case, an outage of the SbW system implies serious effects on people and things. On the other hand, the usage of SbW systems offers a number of advantages such as the elimination of the cross-car steering apparatus, with benefits in terms of greater flexibility in the car design and better passive safety of the driver in case of frontal collisions.

The diagram of a SbW system is drawn in Fig. (**6**). The diagram contains three modules (input, management and actuation) and a communication network. Attention is here paid to the actuation module. The electric motor and the mechanical limbs form the electromechanical actuator. The motor is typically a 3-phase ac machine supplied by a power inverter that is commanded by an Electronic Control Unit (ECU). The latter unit is also in charge of controlling the steering maneuver in a closed-loop fashion thanks to the transducer of the actual steering angle θ_v. The diagram of Fig. (6) is derived from the traditional solution, with the wheels linked by a bar and a single motor moving the bar of a stroke proportional to the steering angle.

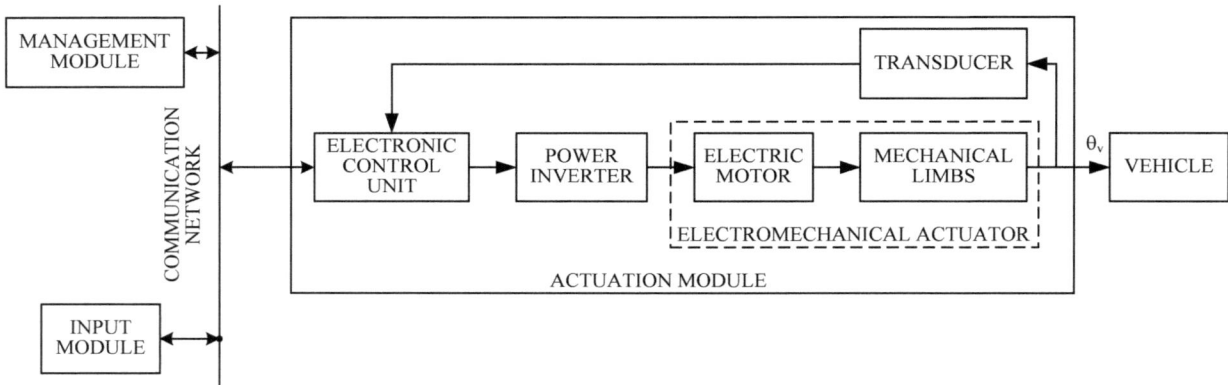

Fig. (6). Diagram of a SbW system.

Let the fault-tolerance level of the SbW system be fail-operational, i.e. the steering maneuver must be executed even if the system fails, and let us consider the case of a failure of the power inverter embedded in the actuation module due to a fault of short-circuit type in one of its switches. To protect the inverter and the connected apparatus, a fuse is inserted in series to each switch of the inverter so that the short-circuit of a switch produces the opening of the half leg containing the faulty switch. Note that the same occurs in case of an accidental blow out of the fuse.

According to the dependability theory, the actuation module is made fail-operational by arranging a redundant inverter and by adopting a fault reconfiguration strategy to tolerate the fault. This can be achieved with the inverter topology of Fig. (7). Besides the traditional three legs, it contains the replica of one inverter leg and three relays. The fault reconfiguration tasks are implemented in the ECU and run at first by detecting the fault, then by locating the fault, and at last by recovering from the fault. The latter task takes two steps: i) isolation of the faulty leg by preventing triggering of the survived switch of the leg, ii) reconfiguration of the motor supply by connecting the replica of the inverter leg to the motor phase.

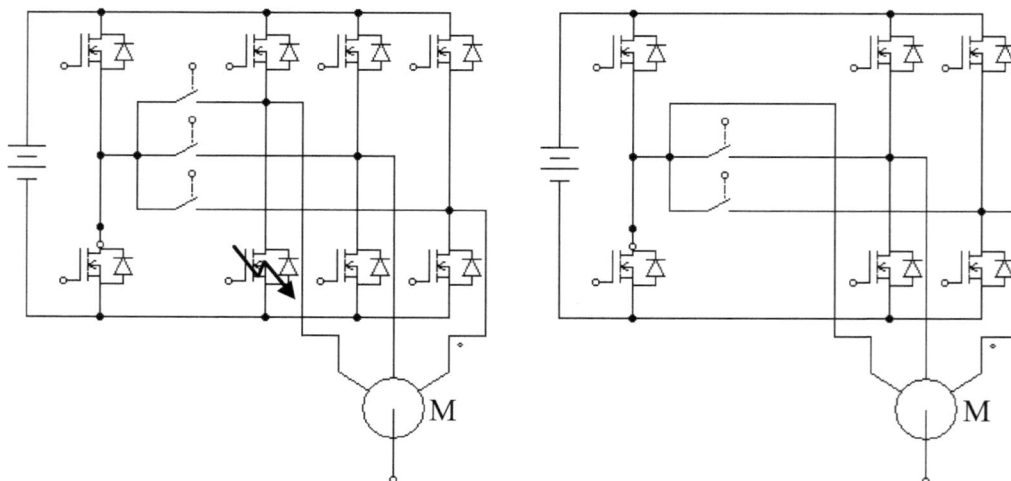

Fig. (7). Fail-operational topology and operation of the power inverter (M denotes the motor).

According to the functional safety standards, the redundant inverter is a safety-related system that performs the safety function of facing up to the fault of a switch. The Risk Matrix technique is adopted to assign the SIL for the redundant inverter. To this end, the Risk Matrix is at first determined by reckoning with the risk probability and the risk severity for an accident of the vehicle, which is the EUC of the case of study. Let the resulting matrix be that one reported in Table **10**. Afterwards the risk ensuing from the fault of a switch is estimated. The risk probability depends on the technical characteristics of the inverter and is assumed to be low. The risk severity depends on the effects of the interruption of the steering maneuver and is assumed to be catastrophic. From Table **10**, it turns out that the SIL of the redundant inverter must be 3.

CONCLUSIONS

The chapter has dealt with the topical subjects of the dependability and the functional safety of a system in the light of the corpuses of the dependability theory and the functional safety standards. After some preliminary definitions on the notion of system and service, the two corpuses have been explicated. Concepts developed to give the matter an abstract but useful assessment have been illustrated. It has been also discussed how the issues treated by the two corpuses integrate each other when dealing with safety-critical systems. In the last paragraph, the theoretical formulations have been put into practice by considering a steer-by-wire system of a vehicle.

DISCLOSURE

A portion of this chapter is based on the authors' past publication in IEEE International Symposium on Diagnostics for Electric Machines, Power Electronics and Drives, 2007 (SDEMPED 2007) titled "A Comparative Explication of the Dependability Theory and the Functional Safety Standards".

REFERENCES

Al-Kuwaiti, M., Kyriakopolous, N., & Hussein, S. (2009). A Comparative Analysis of Network Dependability, Fault-Tolerance, Reliability, Security, and Survivability. *IEEE Communications Surveys & Tutorials, 11*(2), 106-124.

Avizienis, A. (1997, April). Toward Systematic Design of Fault-Tolerant Systems. *Computer, 30*(4), 51-58.

Avizienis, A., Laprie, J.C., & Randell, B. (2001). Fundamental Concepts of Dependability. *Laboratory for Analysis and Architecture of Systems LAAS-CNRS Technical Report*, Apr., 1-145.

Avizienis, A., Laprie, J.C., Randell, B., & Landwehr, C. (2004). Basic Concepts and Taxonomy of Dependable and Secure Computing. *IEEE Transactions on Dependable and Secure Computing, 1*(1), 11-33.

Birolini, A. (2003). *Reliability Engineering. Theory and Practice.* Berlin (Germany): Springer Verlag.

Buja, G., Castellan, S., Menis, R., & Zuccollo, A. (2004). Dependability of Safety-Critical Systems. *Proceedings of IEEE International Conference on Industrial Technology*, CD-2686, 1-6.

Buja, G. & Menis, R. (2007). A comparative explication of the dependability theory and the functional safety standards. *Proc. of 6ᵗʰ IEEE International Symposium on Diagnostics for Electrical Machines, Power Electronics and Drives*, 115-120.

Burcsuk, J. (2007). Development of Safety Related Systems. *Proc. of International Forum on Strategic Technology*, 564-569.

Carey, M. & Purewal, S. (2001). Developing a Framework for Addressing Human Factors in IEC61598. *Proceedings of 2ⁿᵈ International Conference on Human Interfaces in Control Rooms, Cockpits and Command Centres, 481*, 42-47.

IEC TC 65. (1998). *IEC61508-1÷6: Functional Safety of E/E/PE safety-related systems.*

Isermann, R. & Ballé, P. (1997). Trends in the Application of Model-Based Fault-Detection and Diagnosis of Technical Processing. *Control Engineering Practice, 5*(5), 709-719.

Nelson, V.P. (1990, June). Fault-Tolerant Computing: Fundamental Concepts. *Computer, 23*(7), 19-25.

Poledna, S. (1996). *Fault-Tolerant Real-Time Systems; the Problem of Replica Determinism.* Dordrecht (NL): Kluwer Academic Publishers.

Purewal, S. & Waldron, M.A. (2004). Functional Safety in Application of Programmable Devices in Power System Protection and Automation. *Proc. of 8ᵗʰ IEE International Conference on Developments in Power System Protection, 1*, 295-298.

Storey, N. (1996). *Safety-Critical Computer Systems.* Boston: Addison-Wesley Longman Publishing Co.

CHAPTER 4

Steer-by-wire Control System Using GPS for Articulated Vehicles

Rami Nasrallah[1] and Sabri Cetinkunt[2*]

[1]*Caterpillar, Inc., Peoria, IL, USA*

[2]*Department of Mechanical and Industrial Engineering, University of Illinois at Chicago, 842 W. Taylor Street (MC 251), Chicago, IL 60607, USA*

Abstract: Global positioning system (GPS) and embedded computing technology offer new automation opportunities in mobile equipment applications. GPS signals are widely used in direction guidance of cars, ships and airplanes. Construction, mining and agricultural equipment industries stand to benefit from the application of GPS in various control functions. Using GPS and digital embedded control technology, many motion control functions such as motion planning and motion execution can be automated. Automated control of various machine functions without operator (human) involvement brings the embedded control technology closer to autonomous operation capability. Many of these machines require an expert operator who must deal with steering, throttle, and multi-axis tool motion control tasks at the same time, while being very careful in the rather hazardous work environment. It is, therefore, desirable to be able to operate these machines without the operator, that is autonomously, and remove the operator from the dangerous work environment. Autonomous steering of a motor grader is studied in this paper. The system automates the process of generating the desired motion path for a given field operation, and then using GPS signals for its position measurement, it controls the steering system to follow the planned path. This concept can then be integrated with automatic throttle, transmission, brake and tool control sub-systems to achieve completely autonomous mobile machine operation. The system design details of the by-wire-steering system are discussed in hardware and software components. Simulation and tests results on a motor grader are presented. Path tracking accuracy of the by-wire steering system using carrier phase differential GPS (CP-DGPS) position signal was less than 10 *cm* consistently in our tests using a motor grader.

Keywords: GPS, DGPS, CP-DGPS, By-wire-steering, steer-by-wire, electrohydraulics, pump, cylinder, valve, ECM (electronic control module), articulated machine, motor grader, autonomous mobile machines.

INTRODUCTION

Embedded computers and global positioning systems (GPS) are reliable, low cost and widely available technologies (Fig. (**1**)). There is growing demand to automate various functions of, and even autonomous operation of, construction, mining and agricultural machinery in order to reduce operating cost and improve productivity and safety. Presence of human operators in the field, such as in mining and construction, invariably reduces the speed of operations, hence reduce the productivity, due to concerns for human safety. As human operators are removed from the direct operation of these machines, the speed of operation can be increased since there would not be a concern for human safety. In addition, the reliance on expert human operators is eliminated; hence the operational labor cost would be reduced. In short, autonomous operation of construction, mining and agricultural machinery is the next logical phase of the evolution of this industry.

Autonomous operation of construction equipment requires computer controlled sub systems. Specifically the "control of the actuation power" must be electrical, where the electrical control signals are decided by the embedded digital controller and amplified by power amplifier circuits. Therefore, all sub-systems are already in transition from

*Address correspondence to Sabri Cetinkunt: Department of Mechanical and Industrial Engineering, University of Illinois at Chicago, 842 W. Taylor Street (MC 251), Chicago, IL 60607, USA; E-mail: scetin@uic.edu

Sohel Anwar (Ed.)

mechanically or hydro-mechanically controlled to electrical or electro-hydraulically controlled systems. The terminology used here is as follows: if the final control signal which actuates the main control element which controls the main power to the actuator is electrical, then we refer to them as electrical control. An example of such a system is a single stage electrically actuated main valve in a hydraulic circuit where the actuation (movement of the spool of the main valve) is powered by an electrical actuator. Electro-hydraulic control refers to the case where the electrical control signal is amplified to a proportional hydraulic signal which is then used to actuate the main control element. An example of such a system is a high power hydraulic system which uses two stage electro-hydraulic valves, where the first stage uses solenoids and the second stage uses pilot hydraulic control pressure (which is proportional to the solenoid current) that is used to actuate the main valve spool.

In a construction equipment, the following sub systems exist which require operator control decisions,

1. **Engine**: throttle level is controlled by an ECM signal instead of the mechanical mechanism connection between the operator gas pedal and the throttle mechanism or fuel injection system.

2. **Transmission**: manual shifting of the gears versus automatic shifting of the gears using electrically controlled valves which actuate the engagement and disengagement of clutch/brake sets in the transmission.

3. **Brakes**: operator brake pedal actuating a pilot valve which then actuates a main valve to apply brake pressure to the brakes, versus, brake valves (which control the pressure in the brake lines and cylinder) being controlled by an electrically controlled actuator (i.e. a proportional solenoid).

4. **Steering**: a hydraulic cylinder actuates a mechanism to control the steering angle (articulation angle) using a dedicated steering pump hydraulic supply line. The motion of the steering cylinder is controlled by a two-stage electrohydraulic valve which is controlled by the ECM based on the operator's command or autonomous algorithm.

5. **Tool (Implements)**: in general, the implement mechanism such as the cylinders and mechanism to control the bucket of an excavator are powered by hydraulic power. In older generation of designs, the hydraulic power is controlled by two-stage valves where in the first stage; a pilot valve is controlled by the lever movement by the operator (Cetinkunt, 2007). The operator moves the lever which is mechanically connected to the pilot valve and pilot valve outputs a proportional pressure. Then this pressure actuates the main valve which shifts its spool in proportion to that pressure. In digitally controlled systems, the pilot valve is not mechanically connected to the lever, but is actuated by an electrical signal from the ECM. ECM may read the desired pilot pressure from a sensor connected to the lever or from a higher level motion planner.

In this paper, we focus on the by-wire-steering control of a motor grader (Fig. **2**). In order to implement the by-wire-steering system, the steering hydraulic circuits have been modified to have electro-hydraulic two stage valves. The ECM receives the desired steering angle command either from a sensor connected to the steering wheel or joystick if the machine is in manual steering mode, or from a higher level control algorithm which plans the motion of the machine (using GPS position signals for actual position measurement) and steering system commands if the machine is in autonomous mode. Then, the ECM implements the steering tracking algorithm and sends proportional signal to the steering valve. The first stage of the valve accepts the signal from ECM as input and provides a proportional output pressure, which is the pilot control pressure. This pressure then acts on the second stage of the valve (main valve). Since the main stage valve has a centering spring with constant stiffness, the displacement of the spool is proportional to the pilot pressure in steady state. Hence, the steady state speed of the cylinder is proportional to the pilot pressure, or ECM's command signal, assuming constant pressure differential across the main valve, which is the supply (pump output) pressure minus the load pressure. Motor grader has two degrees of freedom of motion for steering (Fig. (**2**)).

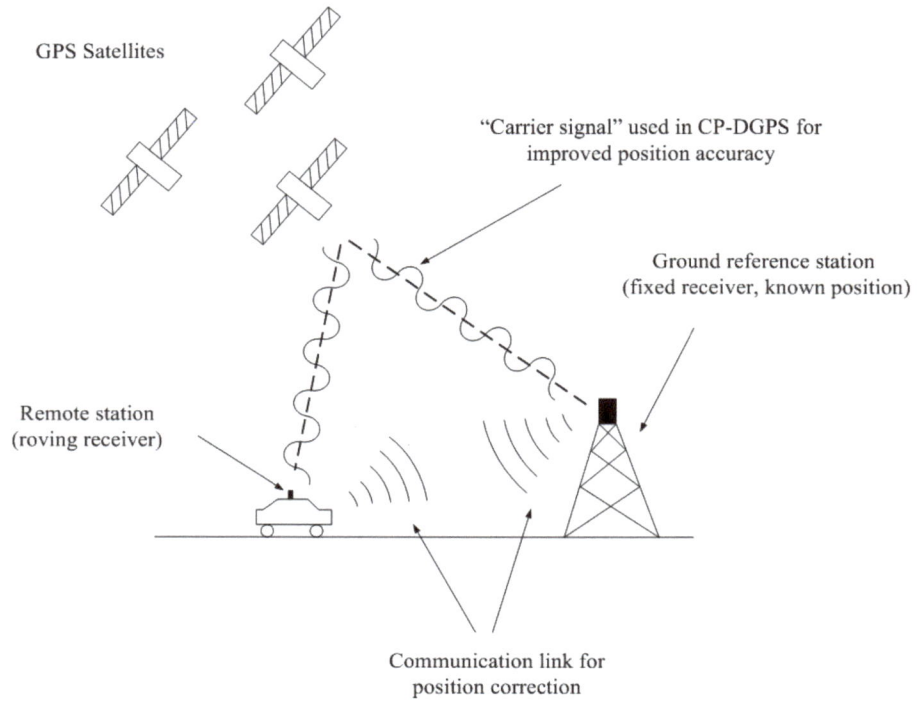

Fig. (1). GPS usage in mobile machine guidance and control.

1. Articulation of the front frame with respect to the rear frame about a hinge joint using a hydraulic cylinder as the actuator

2. Front-wheel steering capability.

Fig. (2). Motor grader and its steering mechanism kinematics.

In addition, the front wheels can be leaned against the sides in order to improve its traction capability. The motion of the blade has six degrees of freedom (DOF). The articulation feature is most useful for cases where sharp turns need to be made, such as when performing a cul-de-sac maneuver. This capability gives the motor grader a relatively low turning radius when both steering and articulation are engaged. The steering and articulation ranges for the same motor grader are approximately 50° and 20° in both directions, respectively. A motor grader is mainly used as a high precision surface finishing tool and requires a skilled operator and complex controls. Some applications of motor graders include soil grading and leveling, soil mixing and creating slopes.

From the operator's perspective, any given construction task can be generalized into three steps:

- Define a path for the vehicle to follow.

- Follow the path using steering (as well as throttle, transmission and brakes).

- Operate the blade mechanism to perform the task.

Automated blade control problem for motor graders was addressed at (Gomm *et al.*, 2009) which uses in-cylinder position feedback sensors and motor grader kinematics. The real time control algorithm also guarantees collision avoidance.

The path of a construction or agricultural vehicle depends on three factors: type of the task, type of the machine and the experience of the operator (Oksanen *et al.,* 2005). Unlike the tasks performed by farm tractors, such as plowing and tilling, motor grader operations usually require that the operator to follow a path such that the soil is always removed to the same direction during the application. The operator would normally mirror the blade about the mirror plane each time the direction of travel of the vehicle is reversed (Gomm & Cetinkunt, 2007).

In motion planning, this type of problem would be referred to as a coverage type problem, where the goal is to define a path which will cover a certain area when traversed (Choset *et al.,* 2005). A classical example is the lawn mowing problem (Arkin *et al.*, 2000), where a desired path is genrated such that the lawn mower would mow the entire lawn, preferably without visiting a certain area more than once. Path planning for farming machinery were studied to get some insight on the algorithms generally used in the field. In (Oksanen *et al.*, 2005), an algorithm is proposed for path planning for a polygonal shaped field, which in the most general case can be concave. One main feature of their method is the use of a trapezoidal decomposition (O'Conner *et al.*, 1995) to split the field into convex shaped polygons and apply a cost function to determine the best direction for the parallel lanes. Similar work in (Jin and Tang, 2006) used Boustrophedon decomposition. For a convex shaped polygon with no obstacles, the path returned by these algorithms would comprise of parallel straight lanes starting at one end of the field and terminating at the other end, with the longest side of the field as the direction for all parallel lanes. In (Reid *et al.*, 2000), it was observed that most path planners for farming machinery are based on guidance of the vehicle in straight lanes, which simplifies the task of the planner.

Turns to connect the end of the current straight lane with the next straight lane are another aspect of path planning. In (Kise *et al.*, 2000) a method is developed for generating turns (Fig. (**3**)) using third-order spline functions. Constraints were set on the minimum turning radius and the maximum steering speed at each generated point on the turn. One drawback of the method is that the number of iterations is high, since a new spline is generated until every point meets the previous constraints. We propose a computationally less expensive method for generating forward turns without using third-order splines.

For any vehicle automatic lateral control system, three main elements need to exist (Fenton *et al.,* 1976):

a) A reference system to generate measurable signals which can be sensed by a vehicle so that its present position can be accurately determined, *i.e.* GPS reference signals.

b) Sensors mounted on the vehicle which measure the signals produced by the reference system and determine how much the vehicle's state should be modified, *i.e.* GPS and ECM.

c)   A steering control system which uses the sensed position so as to maintain the vehicle in a desired lateral position, *i.e.* closed loop controller using GPS.

Points (a) and (b), refer to the technologies which should be used to define the global position of the vehicle, Commercial applications of GPS in guidance are enormous. GPS is formed by a constellation of about 24 orbital satellites used to measure position with respect to these satellites through a method called triangulation, where at least 3 satellites need to be visible to calculate the position at the receiver. In the differential global positioning system (DGPS), ground based reference stations with known positions are used along with the GPS satellites to further reduce the error in position measurement. The carrier phase differential global positioning system (CP-DGPS), which is also known as real time kinematic global positioning system (RTK-GPS), makes use of the high frequency content of the carrier signal of the GPS to obtain higher position accuracies than both GPS and DGPS (Fig. (**1**), Fig. (**4**)).

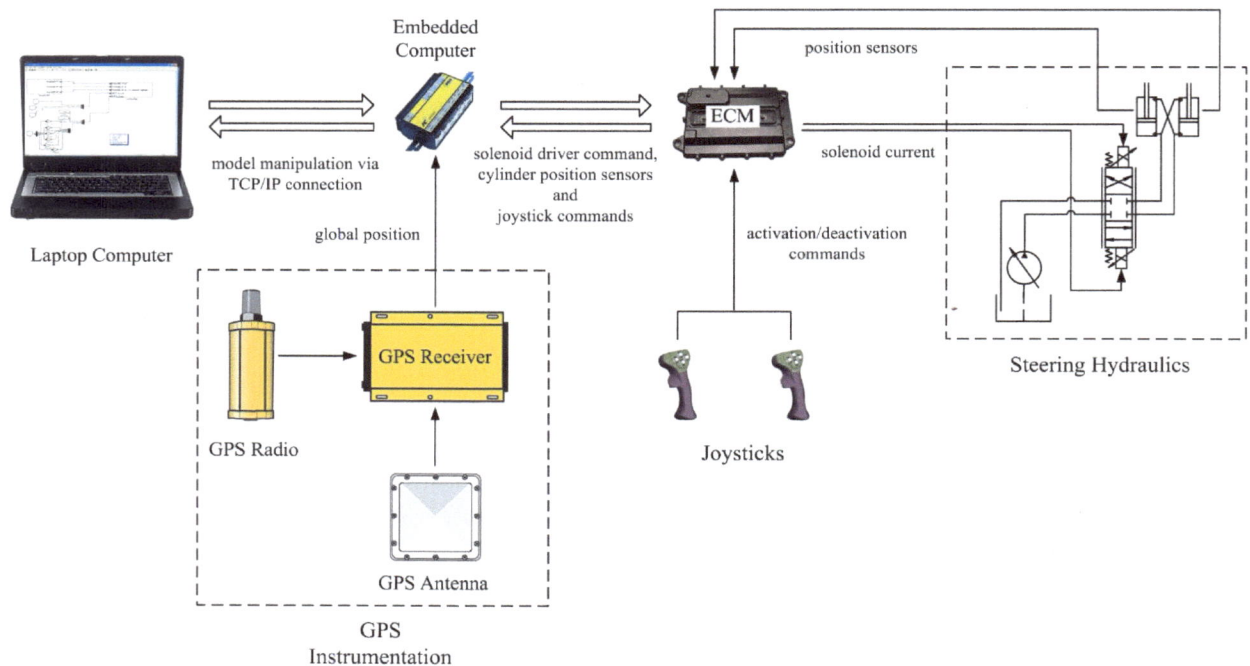

Fig. (3). Control system hardware: ECM, hydraulic circuit, and sensors.

Due to the high accuracy of the CP-DGPS technology, it has been used extensively in path tracking problems. Several authors noted the use of only a single CP-DGPS unit for path tracking, irrespective of the application platform. In our work a single CP-DGPS receiver is used to provide the current global position coordinates of the vehicle.

One of the significant work on path tracking for farm tractors is the work of (Thuilot *et al.*, 2001). The model was approximated by a linear one, which enables the use of linear systems theory. In (Lenain *et al.*, 2004) their work included the effects of wheel slip and predictive control, respectively,

In geometric approach, the pure pursuit method as proposed by (Amidi *et al.*, 1990) and (Howard *et al.*, 2006) have been the most widely used among path tracking algorithms. The method is based on connecting a start and goal point with an arc, the steering command for the vehicle (Fig. (**8**)). The start point is chosen as the tracking point on the vehicle and the goal point is some point at a look-ahead distance away. This distance is similar to the minimum distance the driver would need to react to any change in the road ahead, *i.e.* a change in the curvature. For the pure pursuit method, this is the only parameter which needs to be tuned, and obviously is a function of the vehicle ground speed. Some of the applications for which the method has been used include commercial vehicles tracked mobile robots and a high mobility multi-purpose wheeled vehicle (HMMWV). A similar method is "following the carrot".

The method has two main drawbacks; vehicle oscillations about the path and cutting corners (Makela, 2001) and (Barton, 2001). A recent geometrical method is the vector pursuit method developed by (Wit, 2000). This method differs from the previous two methods in that it takes into account the desired orientation in addition to the position of the vehicle at the goal point. Results by (Wit *et al.,* 2004) suggest that the method is less sensitive than the pure pursuit method to changes in the look-ahead distance. Both vector pursuit and pure pursuit methods give almost similar position and heading errors, with the exception of the fact that the pure pursuit is more sensitive to the look-ahead distance (Lundgren, 2003).

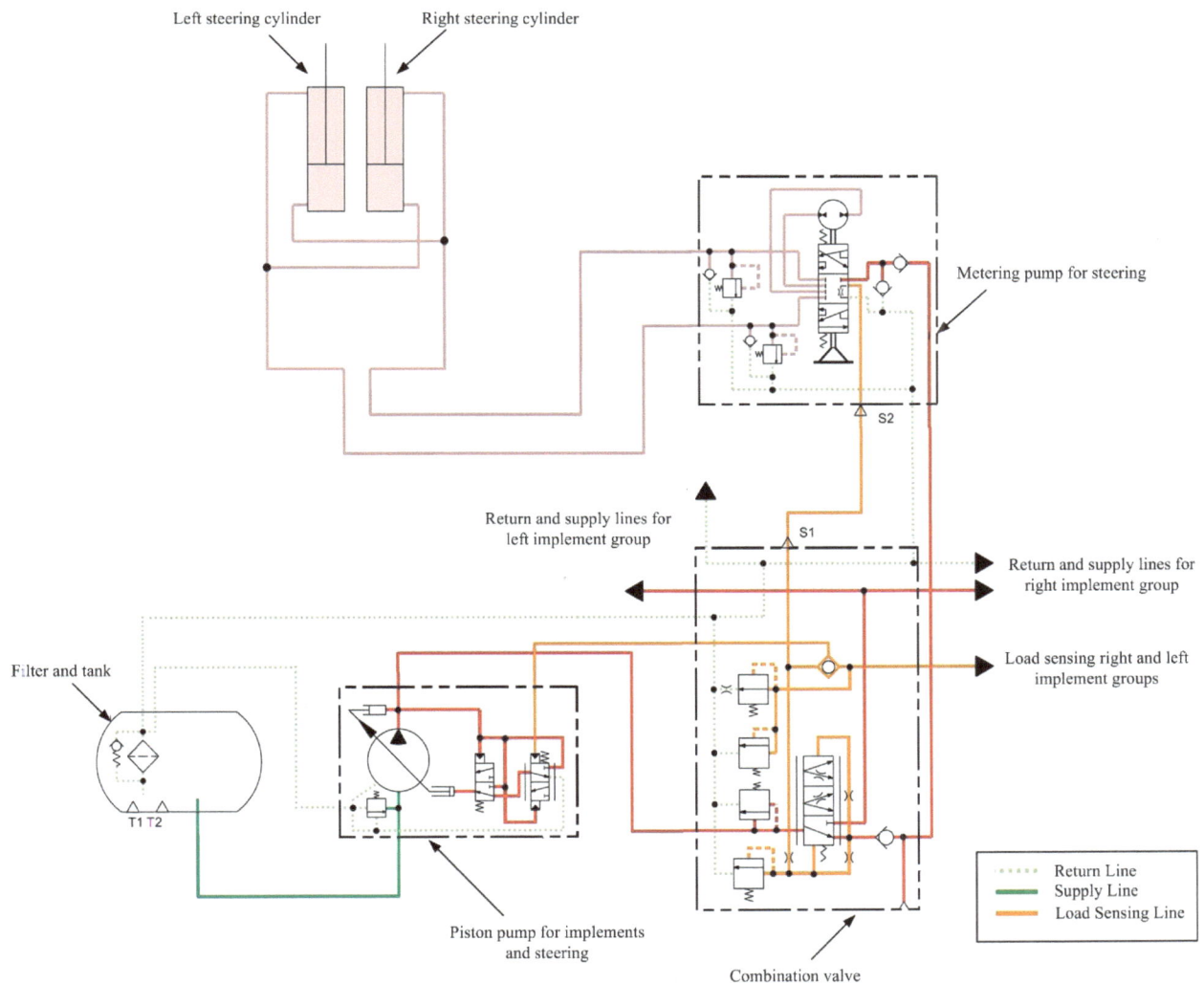

Fig. (4). Electro-hydraulic circuit for the steering system of a motor grader.

STEER-BY-WIRE CONTROL SYSTEM FOR MOTOR GRADERS

Control System Hardware

Figs. (**3** and **4**) show the hydraulic power circuit (pump, valves and cylinders) for the steering system for a motor grader. In addition, it shows the embedded controller (also referred as ECM), GPS receiver and on-machine position sensors for articulation and steering angular position measurement. Due to large hydraulic power levels involved, all valves are two-stage, proportional electro-hydraulic valves, in the modified circuit for the development machine.

Embedded Real-Time Control Code Structure

The path planning algorithm was developed using the C programming language. A dynamic memory allocation strategy (specifically, using singly-linked lists) is used to store the path points. This eliminates specifying an upper limit on the number of points which form the path, as would be the case if arrays were used. Furthermore, manipulations can be easily made to the list, such as adding or removing a point or a set of points to or from the list. In final product implementation of this code, the maximum limit on the dynamic memory allocation should be set in order not to exceed the available physical memory of the embedded controller. There are two major software components to the steer-by-wire system using GPS:

Path Planning: We derive methods for creating paths for the motor grader to follow and cover a given field, provided with minimal information about the field. Due to the computational requirements, the field is represented by a convex shaped polygon with no obstacles within the field.

Path Tracking: The path tracking capability would eliminate the steering and articulation controls from the operator's tasks and perform closed loop steering using GPS position signals.

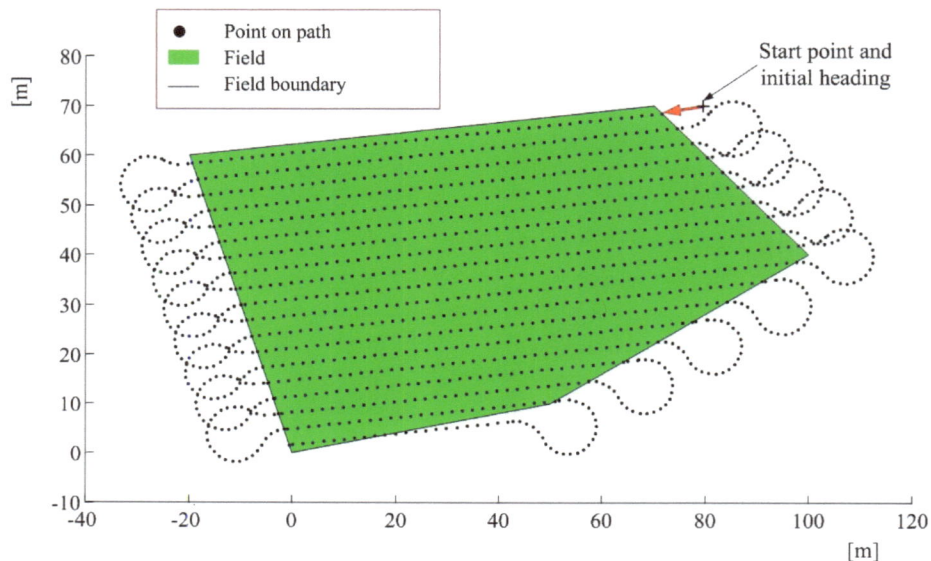

Fig. (5). Path planning and typical path for a motor grader site-work application.

By-Wire Steering System using CP-DGPS

Fig. (5) shows an example of a field with a typical path as returned by the path planning algorithm. From a geometric point of view, each field is represented as a convex polygon while the planned path is a trail of discrete points defined in 2-D Cartesian space. Once the points are connected, they form straight lines and arcs. The straight lines represent the lanes on which the motor grader will travel, while the arcs are used to form the forward turns, which connect each lane with the following one. Several parameters are required to define the complete path:

- Coordinates of the boundries of the field.

- Start point and initial heading of the motor grader.

- Overlap, blade length and the blade circle angle of the motor grader.

The start point and initial heading define the vertex and edge of the field the motor grader will start closest and parallel to, respectively. All lanes are therefore parallel to this starting edge. Information on the overlap, blade length and blade circle angle define the number of forward turns, therefore the number of lanes the motor grader will traverse.

From the assumption that the vehicle can perfectly track a given path, certain conditions need to be met by the path points so as not to conflict with the motion kinematics of the vehicle:

1. **Continuity**: The 1st and 2nd derivatives of displacement with respect to time are defined at every point on the path and are continuous.

2. **Turn Radius**: The radius of curvature at any point on the path is greater or equal to the minimum turning radius of the vehicle used.

Points which lie on the lanes satisfy these conditions immediately. The problem arises at the end of these lanes, where there are connecting forward turns. It is of interest to mention how these turns are generated. In the following methods which we shall present, the forward turn is basically a turn which would start from the last point on the current lane and terminate at a return distance (Δ) away from the first point on the next lane. This distance is required such that the vehicle's main axis is guaranteed to be parallel to the next lane before entering the field.

Three circles turn: For a turn, this method yields a turn as in Figs. (**7** and **8**). Initially, the circles shown in the figure with centers at the points C_1, C_2 and C_3 are set to have radii equal to some constant R_o, which is not smaller than the minimum turning radius of the motor grader. The locations of points C_1, C_3, A, D and E are immediately known, and have the following values with respect to frame T_1:

The distance L is defined through the following relation:

$$L = E_x - A_x + \Delta \tag{1}$$

The value of L can be positive or negative in order to satisfy the condition that the distance between point D and E must be no less than the return distance, Δ. To get the coordinates of points B and C, we need to determine the center of the second circle, C_2. This can be found by solving the following pair of equations, which are defined with respect to frame T_2:

$$(R_{12})^2 = (x - x_1)^2 + (y - y_1)^2 \tag{2}$$

$$(R_{23})^2 = x^2 + y^2 \tag{3}$$

where: $R_{12} = r_1 + r_2$

$$R_{23} = r_2 + r_3$$

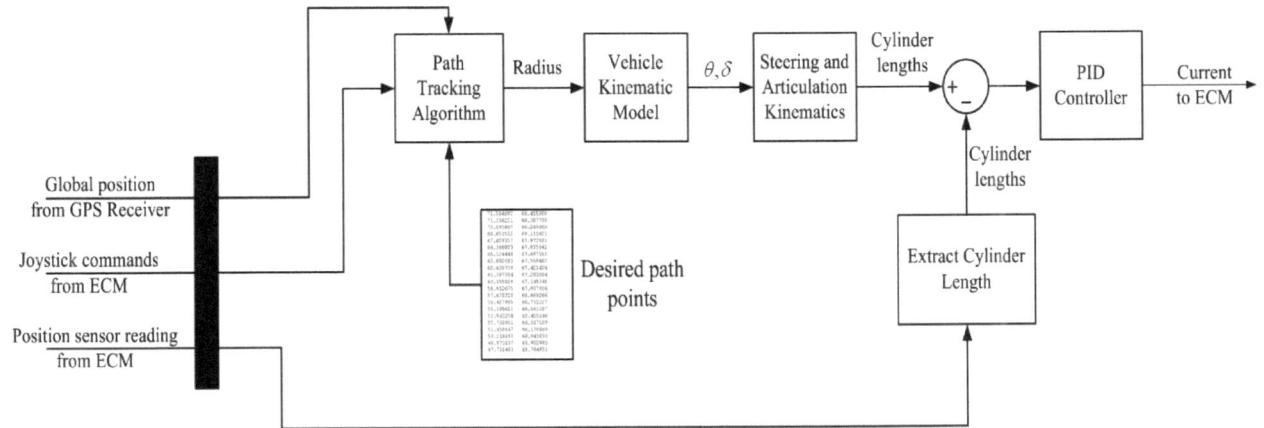

Fig. (6). Real time control algorithm: planning and control for steering.

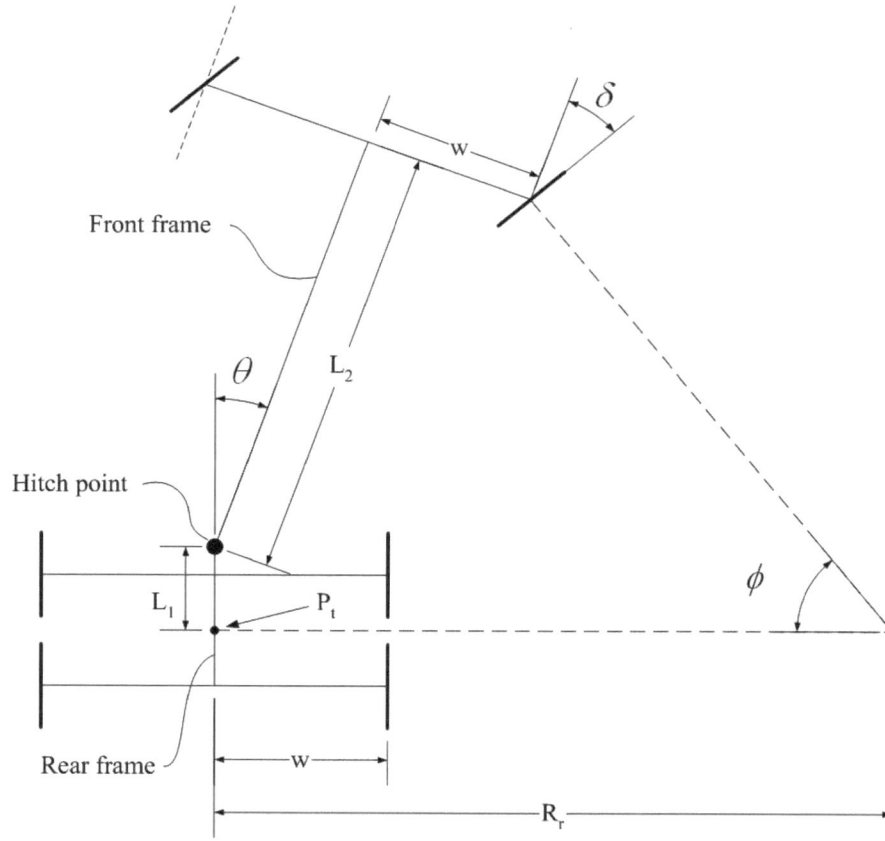

Fig. (7). Kinematics of the motor grader for steering.

2 and 3 define two circles with (center, radius) combination of (C_1, R_{12}) and (C_3, R_{23}), respectively. Point C_2 is the point common to these two circles at one of their intersection points, which are found by finding the roots of the following quadratic equation:

$$Dx^2 + Cx + B = 0 \qquad (4)$$

The roots of equation 4 are in the following form:

$$x_{1,2} = \frac{-C \pm \sqrt{C^2 - 4DB}}{2D} \qquad (5)$$

If the discriminant in equation 5 is negative, then the constant R_o is incremented by some amount and a new set of roots are found again. After finding valid roots, the center of the second circle, C_2, is taken to be located at the root with a larger x-coordinate. The points B and C can be found by determining the angles θ and α shown in Fig. (8).

Two circles turn: Once again for a turn, this method yields a turn as in Fig. (7). Here only two circles are used to form the forward turn. This method is used in cases where the three circles turn method would be undesirable to use, such as in a case when the distance between the last point on the current lane and first point on the next lane is too large.

Parameters H and α in Fig. (7) are found through the following set of equations:

$$H = \sqrt{(2R_o)^2 - d^2} \qquad (6)$$

$$\alpha = \arctan(H/d) \qquad (7)$$

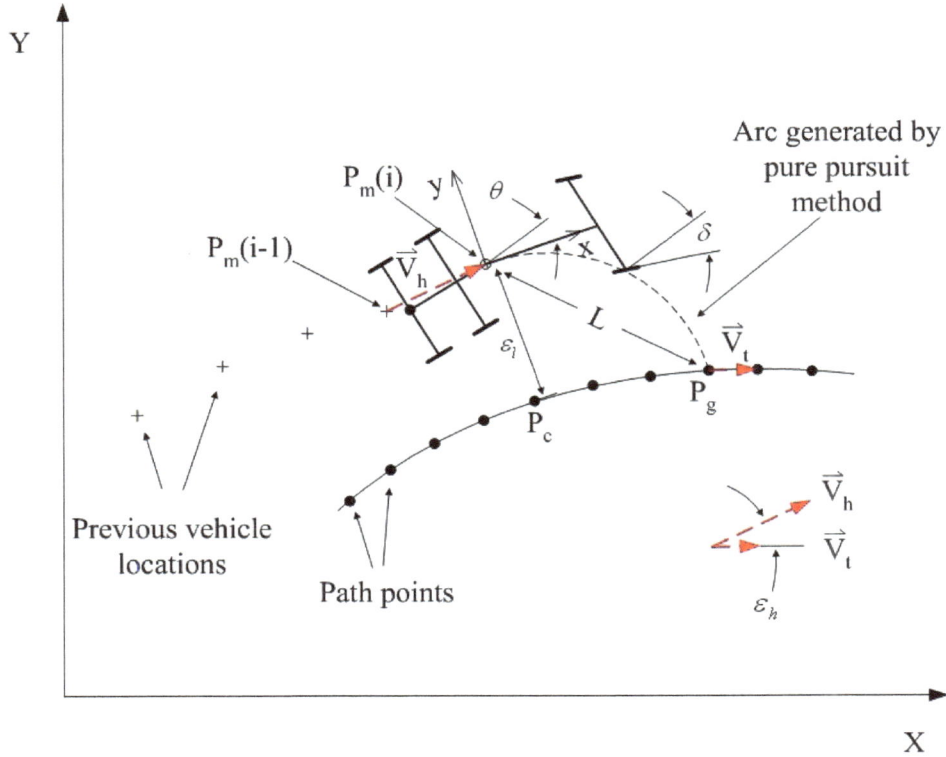

Fig. (8). Path tracking approach: pure pursuit algorithm.

After determining the value of H and α, the location of point B is determined such that the distance L is no less than the distance Δ and is calculated as in equation 1. Parameter h is used to satisfy the condition of having a distance of no less than L between points D and E. This is accomplished by moving point B away from point A, in the direction of the x-axis. The path tracking problem can be viewed as a 2-D problem, where three independent variables (two linear and one angular) are required to uniquely define an object in 2-D Cartesian space. This is shown in Fig. (**8**). Points defined with respect to the machine frame (x, y) are local to the global frame (X, Y). Path tracking commands are defined in the machine frame, while the position of the motor grader is defined in the global frame.

The equation which relates a point defined in the machine frame to its equivalent in the global frame is:

$$\begin{bmatrix} \mathbf{P}_g \\ 1 \end{bmatrix} = \begin{bmatrix} \mathbf{R} & \mathbf{D} \\ 0 & 1 \end{bmatrix} \begin{bmatrix} \mathbf{P}_m \\ 1 \end{bmatrix} \tag{8}$$

Where:

$$\mathbf{P}_g = \begin{bmatrix} P_X \\ P_Y \end{bmatrix}, \quad \mathbf{P}_m = \begin{bmatrix} P_x \\ P_y \end{bmatrix}$$

$$\mathbf{R} = \begin{bmatrix} \cos(\alpha) & -\sin(\alpha) \\ \sin(\alpha) & \cos(\alpha) \end{bmatrix}, \quad \mathbf{D} = \begin{bmatrix} D_x \\ D_y \end{bmatrix}$$

Referring to eqn. 8, $\mathbf{R}$ is the rotation matrix which defines the angle, α, between the machine frame x-axis and global frame x-axis. $\mathbf{D}$ is the translation vector, which locates the origin point of the machine frame, O_m, with respect to the global frame origin, O_g.

Similarly, a point defined in the global frame, P_g, can be transformed to its equivalent in the machine frame P_m, through the following relation:

$$\begin{bmatrix} \mathbf{P}_m \\ 1 \end{bmatrix} = \underbrace{\begin{bmatrix} \mathbf{R}^T & -\mathbf{R}^T\mathbf{D} \\ 0 & 1 \end{bmatrix}}_{\mathbf{T}^{-1}} \begin{bmatrix} \mathbf{P}_g \\ 1 \end{bmatrix} \tag{9}$$

Notice from 8, **T** is the transformation matrix and its inverse, $\mathbf{T}^{-1}$, is shown in equation 9.

The Motor Grader Kinematics for Path Tracking

Figs. (**7**) and (**8**) show the path tracking vehicle model and the relevant parameters for the pure pursuit tracking method.

- $P_m(i)$: Location of the machine frame origin, Pm, at the ith time step.

- $P_m(i-1)$: Location of the machine frame origin, Pm, at the $(i-1)$ time step.

- P_c: Closest point on the path to the machine origin, Pm(i).

- P_g: Current goal point.

- θ: Articulation angle.

- δ: Steering angle.

- $\vec{\mathbf{V}}_h$: Heading vector.

- $\vec{\mathbf{V}}_t$: Tangent vector.

The machine frame is fixed to the motor grader and is defined with its origin, $\mathbf{P}_m$, at the motor grader's hitch point, its x-axis along the front frame of the motor grader, and the y-axis as defined by the right hand rule, such that the z-axis is pointing out of the page. We select our tracking point to be point $\mathbf{P}_m$. Our reasoning for selecting this point is based on the fact that it is close to the blade mechanism center, which would ultimately be the point of interest for following the path. This point also happens to be a good point for placement of the CP-DGPS antenna on the motor grader; therefore we can get the global position coordinates without any transformations. Notice that δ and θ are the control variables for path tracking. The ground speed not considered a control variable. The model is therefore an under-actuated (non-holonomic) one, since two control variables (δ, θ) are used to define three independent variables in 2-D Cartesian space (x, y, α).

The purse pursuit method calculates the radius of curvature of the arc the motor grader would have to travel along, to connect a start point with a goal point. By using this method, we make an assumption that the motor grader tires would interact with the ground in a pure rolling type of motion. Applying the method to the model, the radius of curvature can be obtained through the following equations:

$$R = \frac{L^2}{2x} \tag{10}$$

The parameter L is the look-ahead distance. To apply the method to our tracking model in Fig. (**8**), we connect points $\mathbf{P}_m$ and $\mathbf{P}_g$. A modification needs to be made to eq. 10 to account for the machine frame. Furthermore, we need a sign convention for R to distinguish between a command which steers the motor grader to the left from a command which steers it to the right. Therefore, the radius of curvature for the tracking model is as follows:

$$R = -\frac{y}{|y|}\left(\frac{L^2}{2|x|}\right)$$ (11)

Notice from eq. 11, y is the y-coordinate of the goal point, $\mathbf{P_g}$, as defined with respect to the machine frame. Furthermore, a positive value for R corresponds to a right turn, while a negative one R corresponds to a left turn.

The pure pursuit method returns the desired radius of curvature which the motor grader would have to make; we need to relate this radius to the steering and articulation kinematics of the vehicle. Since there is more than one steering and articulation configuration which can be used for a given radius of curvature for the motor grader, we set a condition such that the vehicle will not articulate in a given direction, only until the full range of steering has been achieved in that same direction. In this way, there is a unique combination for each steering and articulation configuration which can achieve the desired radius of curvature. In practical terms this can be justified since during turns, articulation of the motor grader is necessary only when the steering limit has been reached.

Let us refer to the sum of the steering and articulation angles by the angle ϕ. Equation 12 shows how the steering (δ) and articulation (θ) angles are related. It can be noticed from the equation that articulation is enabled only when steering reaches to its maximum value, δ_{max}, and otherwise articulation is disabled. The sign convention for the steering and articulation angles is positive for a right hand turn and negative for a left hand turn, similar to the convention for the radius of curvature.

$$\phi = \begin{cases} \delta & if\ \delta \prec \delta_{mzx} \\ \delta_{mzx} + \theta & if\ \delta = \delta_{mzx} \end{cases}$$ (12)

To map the radius of curvature, R, to the angle ϕ, which would define the steering and articulation angles, δ and θ, respectively, we construct the kinematic model for the motor grader.

It can be shown that the radius of curvature at the rear frame tandem center (point P_t), R_r, can be related to the steering and articulation angles through

$$R_r = L_1 \sin(\theta) + \xi w \cos(\theta) + \left[\frac{L_1 \cos(\theta) - \xi w \sin(\theta) - L_2}{\tan(\phi)}\right]$$ (13)

where:

$$\xi = -\frac{y}{|y|}$$

Note that ξ carries the sign of the radius of curvature into eq. 13, which would yield the correct sign for steering and articulation angles, depending on the direction of the turn. The radius of curvature of interest is actually at the hitch point; it is related to R_r through:

$$R_h = \sqrt{(R_r)^2 + (L_1)^2}$$ (14)

The pure pursuit method can therefore generate a command to modify the lateral position of the vehicle. There must be some criteria or conditions to determine when to call the method and update the lateral position to eventually track the path. These criteria are as follows:

Heading Error (εh): This is the difference between the vehicle heading, $\vec{V}_h$ and the tangent to the path, $\vec{V}_t$ at the current goal point.

Lateral Error (ε_l): The distance from point $\mathbf{P_m}$ to the closest point on the path, $\mathbf{P_c}$, along the y-axis of machine reference frame during the ith time step.

Distance traveled: To make sure that the vehicle has actually changed its position before modifying the turning radius, the distance between the current location of the vehicle, $\mathbf{P}_m$ (i), and the previous location, $\mathbf{P}_m$ (i − 1), is compared to the arc length between $\mathbf{P}_m$ (i − 1) and the $\mathbf{P}_g$ at the current turning radius.

We combine the pure pursuit method and the previous criteria for updating the radius of curvature of the motor grader.

RESULTS: SIMULATIONS AND EXPERIMENTS ON A MOTOR GRADER

Simulations were necessary in order to verify the methods prior to any testing on actual hardware. A Pentium IV based desktop PC with a Window XP operating system was used for this purpose. The software used includes Matlab and Simulink by Mathworks Inc., and Dynasty1 which is useful for kinematic and dynamic simulations of construction machinery. The basic simulation setup is shown in Fig. (**9**). The joysticks were used to provide the basic inputs from the user to the steering controller such as activation, increasing/decreasing the speed, terminating the simulation, etc.

The control algorithm was implemented using Simulink S-functions written in C language. The same control algorithm is used in simulations as well as on-machine real-time embedded implementations. Using real time workshop (RTW) toolbox, the Simulink model is compiled to another DLL file which can be used to interface it to the dynamic model of the machine (developed in Dynasty). The same Simulink model is converted (using RTW) and compiled it for the target embedded controller to run in real time on the machine and control the machine.

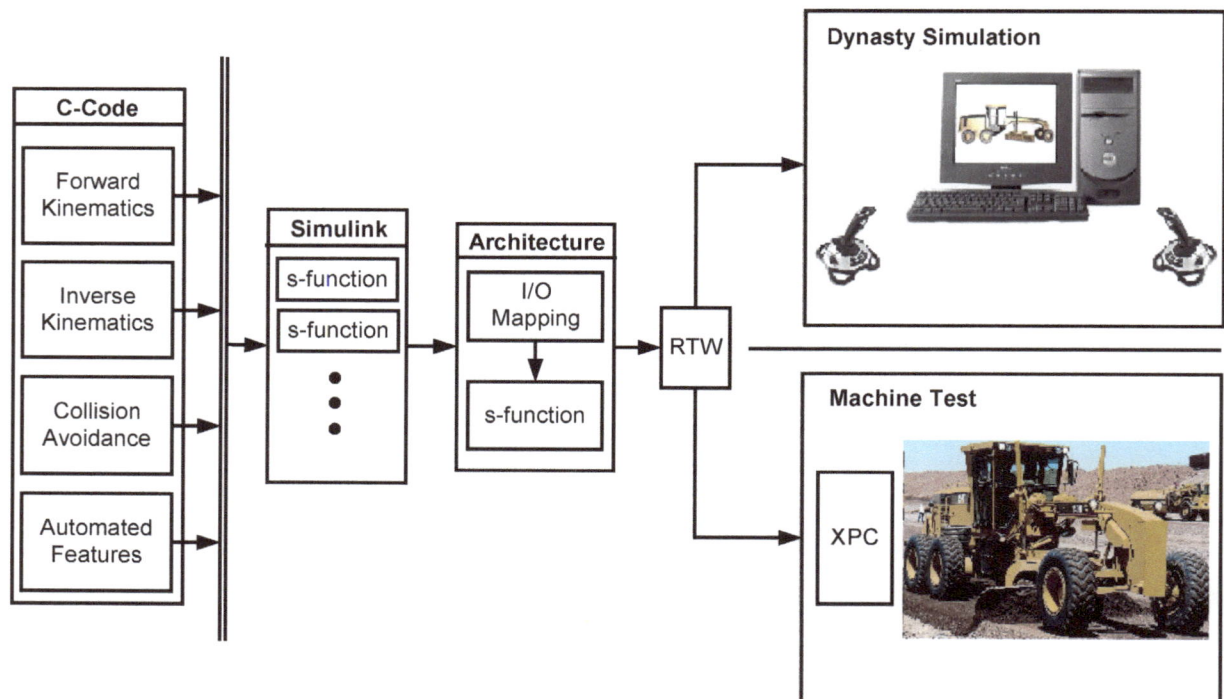

Fig. (9). Control system development environment for simulation and hardware implementation.

Several parameters need to be tuned, both from the path planning and path tracking side, in order that the motor grader track a given path in an acceptable manner. For path planning, these parameters were the distance between points on the path (both on straight lanes and forward turns), the return distance (Δ), and the radius of curvature of the forward turns on the path. As for the path tracking parameters, the only parameter chosen was the look-ahead

---

[1] Dynasty is a proprietary software owned by Caterpillar

distance. Although the ground speed of the motor grader has not been used in developing the path tracking algorithm, several simulation runs showing the effect of various speeds on the path tracking algorithm are presented.

The desired path was comprised of a straight lane followed by a forward turn which connects to another straight lane, thereby reversing the direction of travel of the motor grader.

All simulations were limited to the following conditions:

- The three circle turn algorithm was used to generate the forward turns. The radius of curvature was taken as the minimum radius of curvature of the motor grader, which is approximately 4.88 m. This should not cause a problem, except when the ground speed of the vehicle is too high. To achieve the minimum turning radius, both steering and articulation would have to reach their limits while following the turn.

- The distance between path points was fixed, with a distance no greater than 1 m apart. For regions connecting the straight lanes with turns, the density of the points was greater, since a change in curvature occurs.

- The effect of the ground speed of the motor grader, the return distance and the look-ahead distance were the main parameters of interest for path tracking. Simulation results showing their effect are presented.

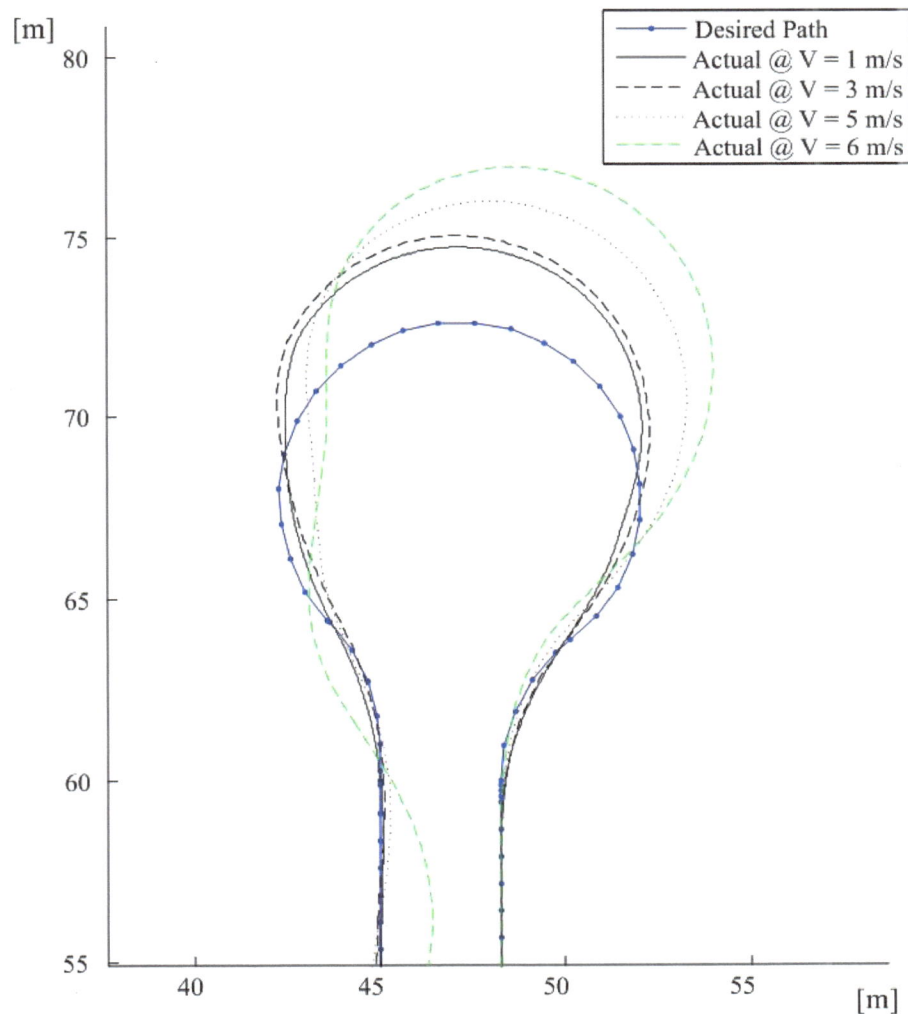

Fig. (10). Effect of travel speed on path tracking accuracy.

To test the effect of the ground speed (travel speed) on path tracking, it was varied from 1 m/s to 6 m/s. Fig. (**10**) shows the simulation results while negotiating a forward turn. For ground speeds up to 4 m/s, the performance of the path tracking algorithm is acceptable, after which the actual paths fail to follow the desired path reliably.

Fig. (**11**) shows the effect of the turning radius on the path tracking accuracy. Notice that the minimum value of this path planning parameter is set by the machine kinematic limits, which is 4.88 m for this particular motor grader.

Fig. (**12**) shows the effect of using different values for the look-ahead distance on path tracking performance. The runs were carried out with a ground speed of 3 m/s and a return distance of $\Delta = 1$m. It can be noticed that small look-ahead values cause a slow response to upcoming changes in the curvature of the path, while large values overestimate the vehicle steering since the change is made too soon. A reasonable value for the look-ahead distance was found to be around 3.5 m. After achieving satisfactory results for path tracking in simulations, the next step was to test the methods on an actual vehicle. Hardware tests were limited to electrohydraulic control of the front wheel steering of the motor grader.

The real-time control algorithm has the following two main components (Fig. (**3**), Fig. (**4**), and Fig. (**9**))

- High Level Controls: These are commands which control the behavior of the motor grader globally, i.e. its global position with respect to the path. For a given test, the Matlab/Simulink model containing the steering controller algorithms was loaded onto the embedded computer in the form of a Dynamic Link Library (DLL) file *via* a TCP/IP connection between the laptop and embedded computer. All necessary operations such as activating, monitoring and deactivating the model were made through the laptop. The joysticks were used to enable the path tracking function of the steering controller and therefore disable any manual steering commands. Once path tracking is enabled, steering commands are generated in the embedded computer after comparing the present location of the motor grader returned from the GPS receiver, to where it should be on the path. These commands are then sent to the steering Electronic Control Module (ECM) to command the steering cylinders. The update time used for the steering controller model was 0.01 seconds that is the sampling rate was 100Hz.

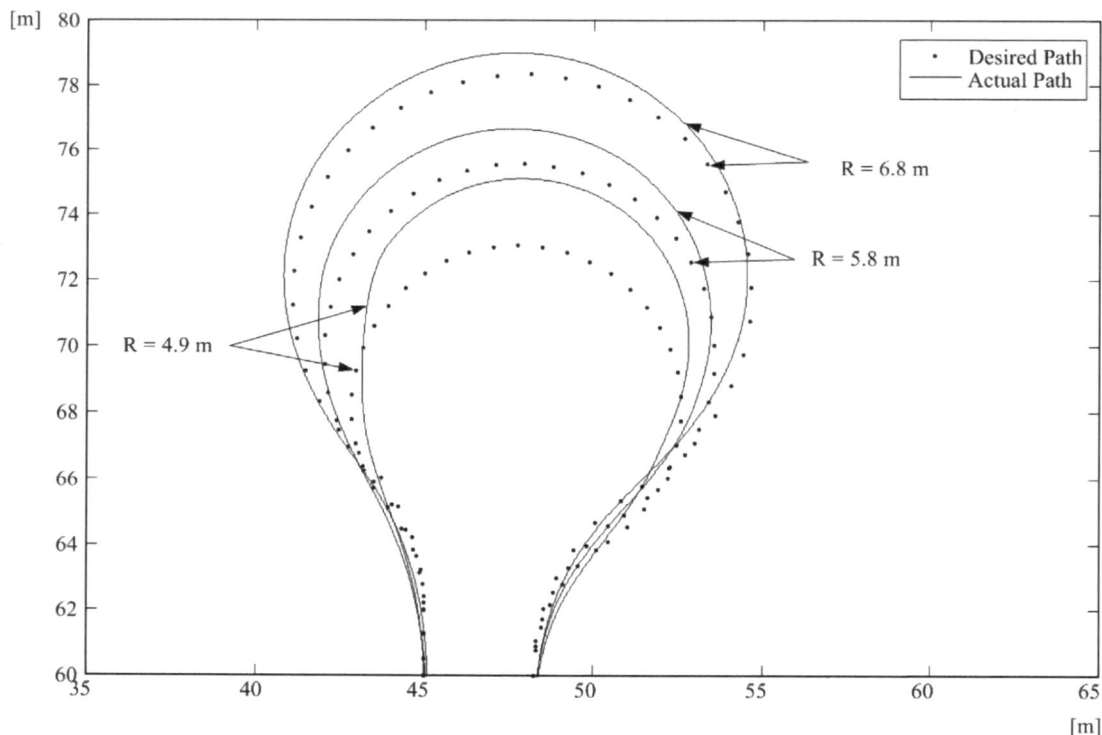

Fig. (11). Effect of turning radius on the path tracking accuracy.

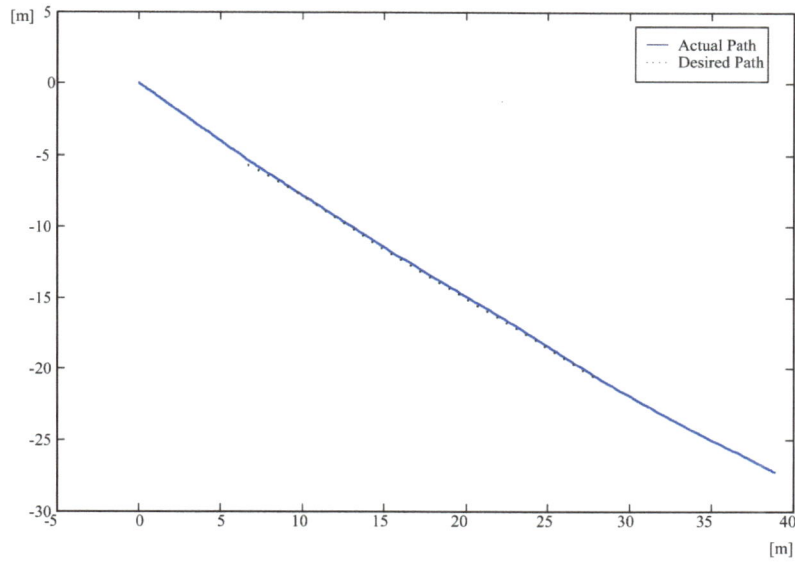

Fig. (12). Hardware test result for a straight line using a motor grader.

- Low Level Controls: These are control commands sent to the front wheel steering actuators. Commands from the embedded computer are passed through the ECM to the solenoid valves which control the front wheel steering cylinders. The commands sent from the embedded computer are in the form of an electrical current which, after amplification in the ECM, can be used to move the hydraulic cylinder. The position to which the cylinder needs to be extended or retracted is determined through closed loop feedback of the position sensors using a PID controller.

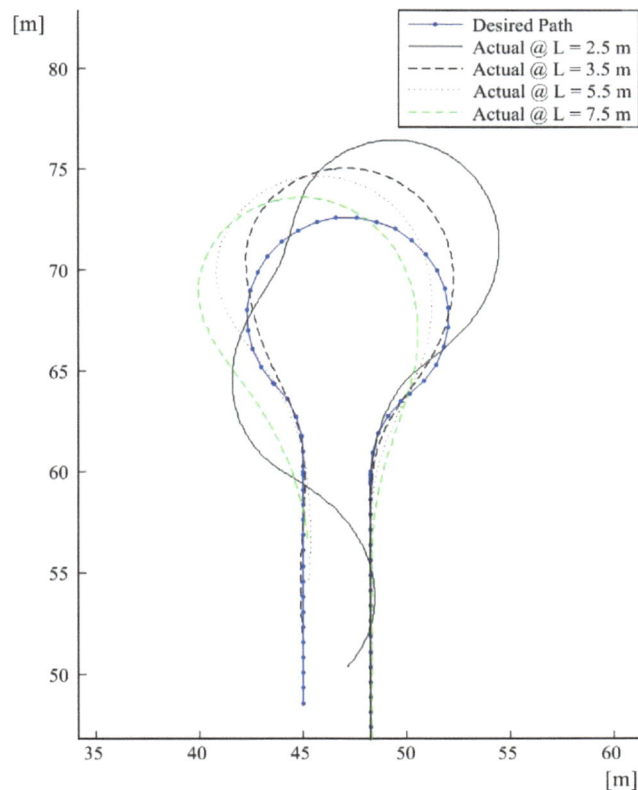

Fig. (13). Effect of different values of look-ahead distance on path tracking accuracy.

Fig. (**6**) shows a block diagram description of the Simulink model which was used for the hardware tests. Notice that the path tracking is enabled or disabled through the joystick signals. Once enabled, the motor grader front wheel steering is eliminated from the operator's functions and is controlled through the steering controller algorithms. Although the controller has been simulated and verified for both front wheel steering and articulation control, hardware testing only involved front wheel steering control.

We performed a couple of tests to check the performance of the steering controller on a straight lined path and a curved shaped path. Notice that for the curved path, the motor grader required steering of the front wheels and no articulation, since the turning radius of the curve was not set to the minimum turning radius of the motor grader. For both tests, the following procedures were carried out:

1. Define the desired path by driving the motor grader manually and capturing the GPS location of the vehicle at time intervals approximately 1 second apart.

2. Return the vehicle to the start point or an approximate location to the first point on the desired path.

3. Prepare the model by downloading the steering controller model onto the embedded computer after it has been compiled with the desired path points included in it.

4. Activate the steering controller to begin path tracking.

Note that data acquisition was incorporated within the model in order to capture the actual paths which the vehicle created and compare with the desired paths.

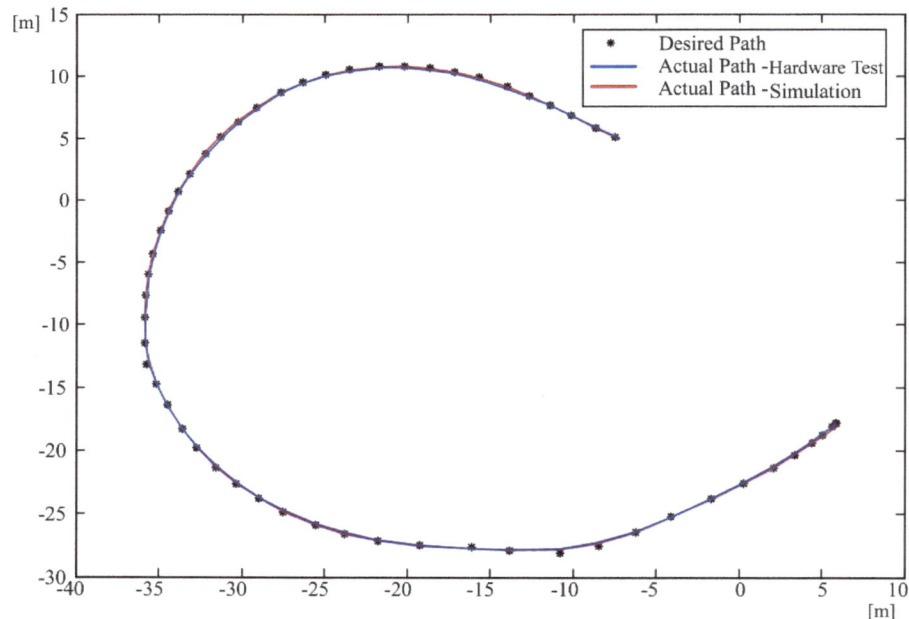

Fig. (14). Comparison of simulation and hardware tests on the motor grader for a curved path.

Simulation and actual hardware test run shown in Figs. (**13** and **14**) show the path tracking results for the motor grader for a straight line travel and a curved path travel. The ground speed was set at 3 m/s.

The absolute path following error for the straight path was 0.0561 m, while for the curved path it was 0.0683 m. For the absolute error calculation of the curved path, we did not take in to consideration the first nine (9) points on the desired path, where the absolute error for these points was higher than the average. This is because the vehicle was initially off-set from the desired path.

CONCLUSIONS

A by-wire steering control system hardware component design and embedded software is presented. The real-time control software development is done completely in Matlab/Simlink environment using S-functions coded in C and Real-time Workshop tools. The identical control algorithm is tested in both simulations with a dynamic model of the motor grader and on the actual machine. Both simulation and experimental results show the feasibility of autonomous steering of a motor grader using CP-DGPS position signals for closed loop autonomous steering of the machine. The next step would be to integrate this capability with automatic throttle, brake and tool control sub-systems in order to realize a completely autonomous machine capability.

REFERENCES

Amidi, O. & Thorpe, C. (1990). Integrated Mobile Robot Control. *The Robotics Institute, Carnegie Mellon University*, May, CMU-RI-TR-90-17.

Arkin, E., Fekete, S., & Mitchell, J. (2000). Approximation algorithms for lawn mowing and milling. *Computational Geometry*, *17*, 25-50.

Barton, M.J. (2001). *Controller Development and Implementation for Path Planning and Following in an Autonomous Urban Vehicle*. Bachelor's Thesis, School of Aerospace, Mechanical and Mechatronic Engineering, The University of Sydney.

Cetinkunt, S. (2007). *Mechatronics*. John Wiley and Sons Inc.

Fenton, R.E., Melocik, G.C., & Olson, K.W. (1976). On the Steering of Automated Vehicles: Theory and Experiment. *IEEE Transactions on Automatic Control*, *21*(3), 306-315.

Gomm R., Bhaskar V., & Cetinkunt S. (2009). Automated Real-Time Motion Planning and Control of Construction Equipment Mechanism. *International Journal of Robotics and Automation*, *21*(1), 143-151.

Gomm, R. & Cetinkunt, S. (2007). Memory Efficient Real-Time Motion Planning by Dual-Resolution Heuristic Search. *Journal of Robotics and Mechatronics*, *19*(1), 114-123.

Howard, T., Knepper, R.A., & Kelly, A. (2006). Constrained Optimization Path Following of Wheeled Robots in Natural Terrain. *Proceedings of the 10th International Symposium on Experimental Robotics* 2006 (ISER '06).

Jin, J. & Tang, L. (2006). Optimal Path Planning for Arable Farming. *Proceedings of the ASAE Annual International Meeting*, 061158

Kise, M., Noguchi, M., Ishii, K., & Terao, H. (2000). Enhancement of Turning Accuracy by Path Plan-ning for Robot Tractor. *Proceedings of the Conference on Automation Technology for O -Road Equipment*, 398-404.

Lenain, R., Thuilot, B., Cariou, C., & Martinet, P. (2004). Adaptive and predictive non linear control for sliding vehicle guidance: Application to trajectory tracking of farm vehicles relying on a single RTK GPS. *Proceedings of the IEEE/RSJ International Conference on Intelligent Robots and Systems*, 1, 455-460.

Makela, H. (2001). *Outdoor Navigation of Mobile Robot*. Doctoral Dissertation, Helsinki University of Technology.

O'Connor, M., Bell, T., Elkaim, G., & Parkinson, B.W. (1995). Kinematic GPS for Closed-Loop Control of Farm and Construction Vehicles. *Proceeding of the Institute of Navigation*, 1261-1268.

Oksanen, T., Kosonen, S., & Visala, A. (2005). Path Planning Algorithm for Field Tra c. *Proceedings of the ASAE Annual International Meeting*, 053087.

Reid, J., Zhang, Q., Noguchi, N., & Dickson, M. (2000). Agricultural automatic guidance research in North America. *Computers and Electronics in Agriculture*, *25*, 155-167.

Thuilot, B., Cariou, C., Cordesses, L., & Martinet, P. (2001). Automatic Guidance of a Farm Tractor Along Curved Paths, Using a Unique CP-DGPS. *Proceedings of the International Conference on Robotics and Automation*, 2, 674-679.

Trimble Navigation Limited. *GPS Tutorial*. Trimble Home Page retrieved 2009, from http://www.trimble.com/gps/index.shtml.

Wit, J. & Crane III, C.D. (2004). Autonomous Ground Vehicle Path Tracking. *Journal of Robotic Systems*, *21*(8), 439-449.

CHAPTER 5

Virtual Operator Model for Construction Equipment Design

Ahmed Adel Elezaby[1*] and Sabri Cetinkunt[2*]

[1]University of Illinois at Chicago, 842 W. Taylor Street (MC 251), Chicago, IL 60607, USA

[2]Department of Mechanical and Industrial Engineering, University of Illinois at Chicago, 842 W. Taylor Street (MC 251), Chicago, IL 60607, USA

Abstract: Developing new construction equipment or modifying an old model requires large investment in re-engineering and testing. This paper presents a virtual operator model that can be used to (1) control, test and evaluate virtual construction equipment before building the prototype, (2) autonomously control the construction equipment after being built. Furthermore, the virtual operator model is adaptive to properly control different models of the same machine. The operator model is useful in evaluating a new machine design and predicting its performance in applications under realistic conditions as a real human operator would operate it. Using this model, potential problems can be identified in early design stage, hence reducing the costly prototype testing stage of the machine development.

Keywords: Virtual Operator Model, Autonomous Control, Construction Equipment, Virtual Product Development, Truck Loading Cycle, Neural Network Controller.

INTRODUCTION

Developing new construction equipment requires large investment in re-engineering and testing (Fig. **1**). After the design is completed, a prototype is built. This prototype is tested by experienced operators through pre-defined tests and performs normal operations of the construction equipment. Usually, many modifications are needed on the prototype machine based on the tests results. The prototype is then modified and tested again. This iterative design and testing process continues until the product meets all performance specifications and confirmed by extensive test results. This procedure requires spending a lot of time and money in building prototypes and testing it. Virtual product development (VPD) has become essential now in developing construction equipments to improve the machine quality, reliability and operability saving significant cost and time in building prototypes and testing them.

Construction equipment operations quality and fuel efficiency depends to a great extent on the operator experience. Every operator does the operations depending on his level of skill and experience. Moving towards autonomous construction equipment operations will standardize operations and will take advantage of the maximum power and energy efficiency of the machine.

The objective is to build a virtual operator model that can (1) control, test and evaluate virtual construction equipment before building the prototype, (2) autonomously control the construction equipment after being built.

Filla *et al.* (2005) presented the results of a simulation model of a human operator that selects simple functions like lift, tilt, throttle, gear, or brake based on general rules that together describe the machine's working cycle. In (Filla, 2005) the operator selects these tasks as a function of events. An example of using dynamic simulation of construction equipment in development was presented in (Filla & Palmberg, 2003). In (Cetinkunt, 2006) the control

*Address correspondence to Ahmed Adel Elezaby and Sabri Cetinkunt: University of Illinois at Chicago, 842 W. Taylor Street (MC 251), Chicago, IL 60607, USA; E-mail: aeleza2@uic.edu; Department of Mechanical and Industrial Engineering, University of Illinois at Chicago, 842 W. Taylor Street (MC 251), Chicago, IL 60607, USA; E-mail: scetin@uic.edu

of ideas, tuning of PID controllers and control of hydraulic systems for construction equipment was described explained. In (Vogel, 2002) the driver behavior was modeled in traffic simulation and is relevant to this work. Other related work are presented in the rest of the references (Grant 1996, Lee et al. 2004, Macadam 2003, Shi 1996, Singh 1995 & 2002).

Fig. (1). Examples of construction equipments.

In the work presented here similar to (Filla 2003, 2005), an operator model selects subtasks as dig and load the truck using an event based strategy model and delivers the human commands to a separate machine model which simulates the wheel loader. The machine model gives the operator model a feedback with data that can be perceived by human operator. The model can modify and tune itself to control different sizes of the machine.

BACKGROUND

The machine used in this research is a wheel loader; it has three main sizes small, medium and large depending on the capacity. One of the main applications of the wheel loaders is truck loading. The operator controls are tilt and lift of the bucket, throttle, steer, gear (F1-F4, N, R1-R4), and brakes.

The Truck Loading Cycle

A typical truck loading cycle can be described by the following events as shown in Fig. (**2**). The wheel type loader, or also called as just "Loader" starts at the truck with the bucket fully lifted and dumped the load in the truck. The loader then backs from the truck while racking back and lowering the bucket. Then, the loader approaches the pile and digs. After digging is done the loader backs from the pile then moves towards the truck while lifting the bucket, then dumps the load in the truck. These events are repeated until the truck is filled with payload.

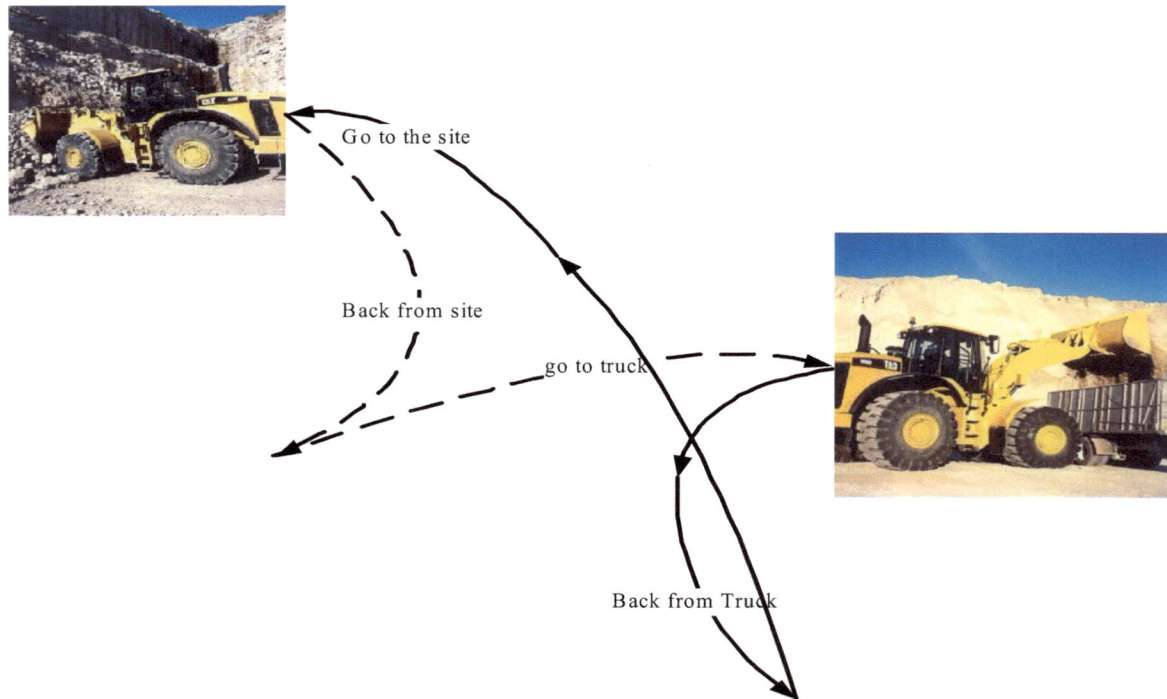

Fig. (2). Truck loading cycle.

The Evaluation Tests

Five evaluation tests were used to determine the performance and the controllability of the wheel loaders and they are explained briefly in this section.

a) The First Test: Raise to a Point

The objective is to quantify the ability to quickly and accurately raise a load to a given height. The machine setup is full throttle, neutral transmission, parking brake applied, racked bucket (angular position of the bucked all the way back towards the vehicle body), at rest slightly above ground line and the response measurements are time from start to end point (sec) and position error relative to target point (mm). This is shown in Fig. (**3**).

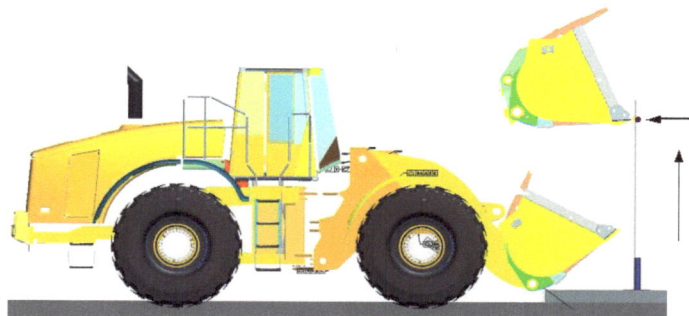

Fig. (3). Test 1 raise to a point.

b) The Second Test: Lower to a Point

The objective is to quantify the ability to quickly and accurately lower a load to a given height above ground line. The machine setup: full throttle, neutral, parking brake ON, racked bucket, at rest near full height. The response measurements are time from start to end point (sec) and position error relative to target point (mm). This is illustrated in Fig. (**4**).

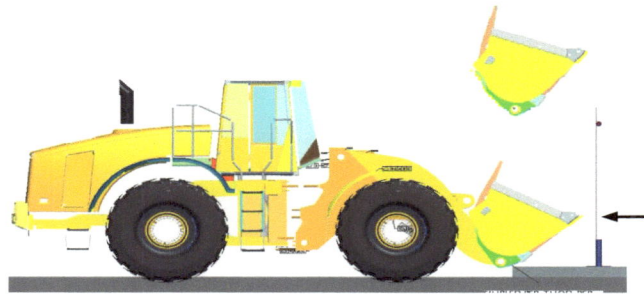

Fig. (4). Test 2 lower to a point.

Fig. (5). Test 3 curl bucket at a given height.

c) The Third Test: Curl Bucket at a given Height

An illustration of this test is shown in Fig. (**5**). The objective is to quantify the ability to curl the bucket cutting edge at a given target height. The machine setup: full throttle, neutral, parking brake ON, racked bucket, lift slightly above level arms, at rest aligned with target height. The response measurements are time from start to full dump (sec) , time for return, full dump to full rack (sec), and position error (mm).

d) The Fourth Test: Vertical Line Test

The objective is to quantify the ability to quickly and accurately control the bucket cutting edge to a vertical straight line. The machine setup: full throttle, neutral, parking brake ON, level bucket, at rest slightly above ground-line. The response measurements are time from start to upper end point (sec), sum position error up = (mm), time = from upper end point to ground-line stop (sec) and sum position error down. This test is illustrated in Fig. (**6**).

Fig. (6). Test 4 vertical line test.

d) The Fifth Test: Follow a String Test

An illustration of the test is captured in Fig. (**7**). The objective is to quantify the ability to control the bucket cutting edge to follow a taught string line as accurately and quickly as possible while traveling rough terrain. The Setup: 1st gear, forward, carry position, full rack-back, align cutting edge with string both start and end. The response measurements are time start to full dump (sec) and sum position error (mm).

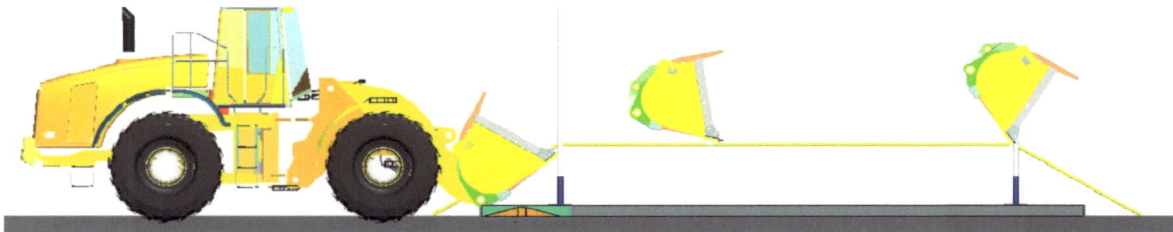

Fig. (7). Test 5 follow a string test.

THE VIRTUAL OPERATOR MODEL

Actual human operator behavior in the recorded data of testing and normal operations was analyzed with a software package which is specialized in interpreting human behavior. The purpose of this analysis was to determine the operator decisions while operating the machine. It was found from the analysis that wheel loader operations can be divided into nine tasks which are the following: lift to a point, lower to a point, dump the load, rack the bucket, backing from the truck, approaching the site, filling the bucket, backing from the site and finally going to the truck. All the wheel loader operations use these tasks to accomplish the selected mission either in series or parallel. The aim is now to model these tasks then design a higher high level supervisory control algorithm to control these tasks either in parallel or series to accomplish the tasks defined in the background section.

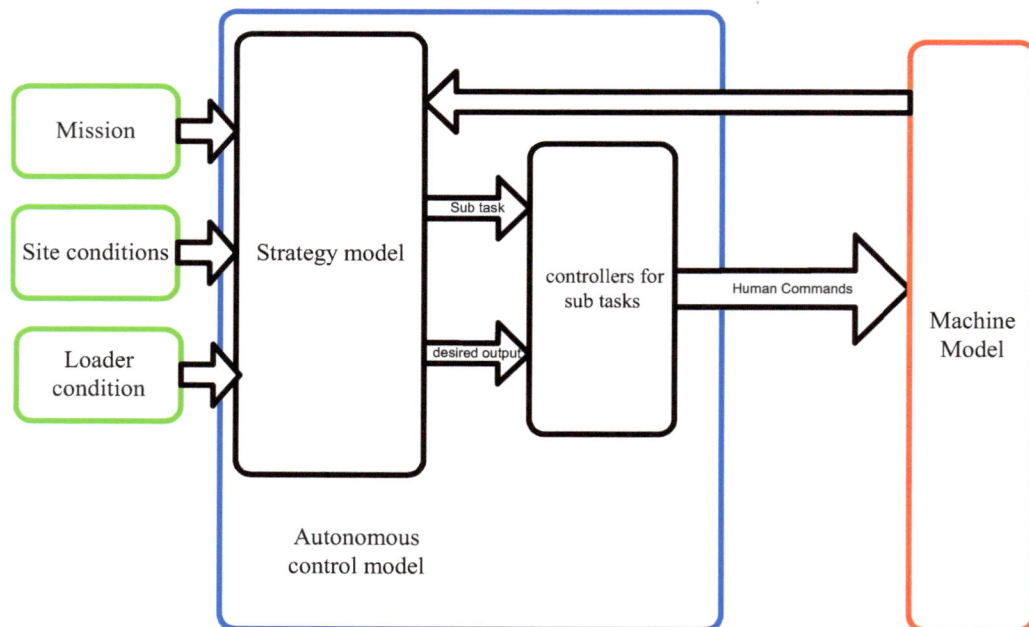

Fig. (8). Virtual model schematic.

The virtual operator model schematic as shown in Fig. (**8**) consists of two main parts: the strategy model and controllers for sub tasks. The inputs of the operator model can be categorized into three categories which are the objective (or mission), site conditions and loader conditions. The output to the machine model are the human commands Real time simulated work-site and machine conditions are fed-back to the operator model and used in the decision making process, as is the case in human operated machines.

The mission is the first category of inputs. The mission input determine the task that should be accomplished by the wheel loader, i.e. select from a complete truck loading cycle until the truck is full or a certain pre defined volume of load that are loaded to truck(s), part of a truck loading cycle, or to do one of the evaluation tasks defined in the background section. The second category of inputs is the site conditions, which defines the position of the truck, the position of the dig site. The height of the truck and the available space in the site for the wheel loader to travel between the dig site and the truck. The third category is the loader conditions which define the position of the wheel loader with respect to the start point of the site at the beginning, the position of the bucket with respect to wheel loader, the volume of load in the bucket, and the steering angle and the direction of the vehicle. All of this sensory information can be made available in a future construction or mining site where all vehicles are networked, have GPS receivers, and machine based actuator position sensors.

The outputs of the operator model resembles the outputs of the human operator to control the wheel loader, Simply are the inputs to machine model or the machine itself which are throttle command input percentage from 0 to 100%, steering wheel angle, the gear shift, the brake command input percentage from 0 to 100 %, the lift and tilt of the bucket command input from -100% to 100 % where from -100 % to 0 for lower the bucket in the case of the lift command and tilt in for the tilt command, and from 0 to 100 % for raising the bucket in the case of the lift command and tilt out for the tilt command.

The feedback from the wheel loader to the operator model is the information that can be perceived by the human operator either through gauges reading in the cabin or by his vision which are the position of the bucket the current velocity of the wheel loader, the engine speed the displacement or the distance moved by the wheel loader, the distance between the wheel loader and the target (either the truck or the dig site), the steering angle , the direction of the wheel loader, the volume of the load in the bucket , the speed of the tires in revolution per minute to determine wheel slippage and the position of the external points of the wheel loader.

The Strategy Model

The strategy model (schematic is shown in Fig. (**9**)) has the highest level of control in the operator model. It is a finite state machine model done in a state flow environment. The strategy model receives the inputs for the operator model which are the mission, site conditions and loader conditions defined in the previous section. It also receives the feedback from the machine model. Depending on this information, it decides the sequence of the sub tasks needed to accomplish the required mission, and selects the sub tasks to be done now either in parallel or series. It can also decide if the vehicle is stable or not and if it is an emergency situation or not and selects the subtask which can deal with current situation. It also determines which controllers of subtasks will be used, and the desired output of the actuator. It continuously calculates the error between the desired output and the current output and decides when to continue to use the same controller or switch to another one.

First after selecting the mission, if it is a complete truck loading cycle or part of it, the strategy model selects the main finite state machine model. If it is one of the evaluation tests defined in the background section, the strategy mode selects another finite state machine models embedded in the strategy model which are more suitable for the evaluation tests. Depending on the site conditions and the loader conditions the strategy model selects the tasks to put the wheel loader in a start state and position to accomplish the mission required. It then selects a task from the following and the sequence needed and the tasks needed to be done in parallel and the ones needed to be done in series.

The tasks are as follows:

1.   Shut down/Emergency Brake or stop

2.   Back from the truck/site

3.   Go to truck/Site

4.   Lift/lower to a point

5.   Dump load/Rack bucket

6.   Gear selector

7.   Filling the bucket

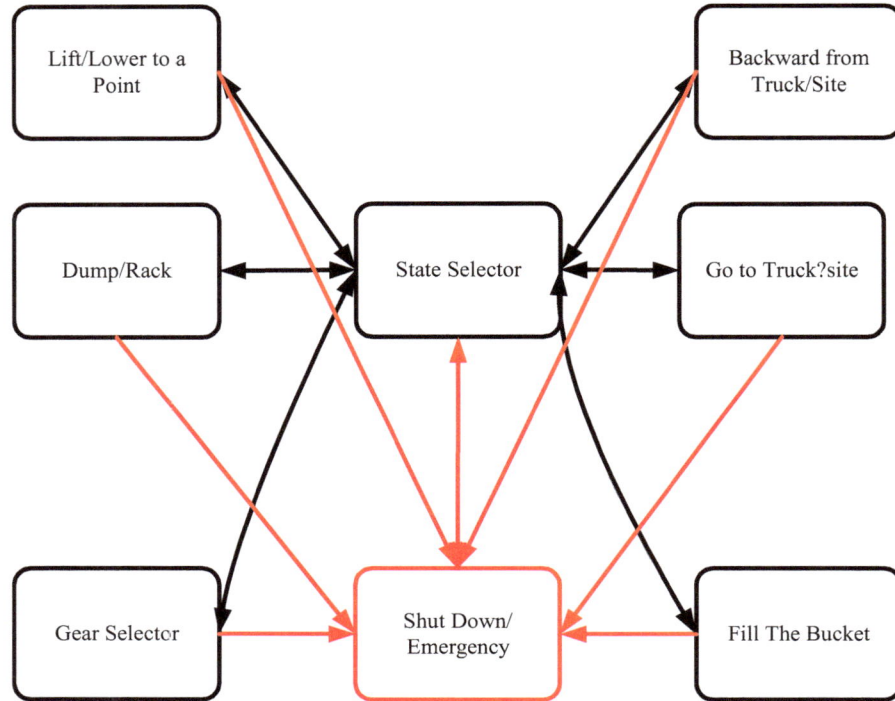

Fig. (9). The Strategy Model Schematic.

The Strategy Model Tasks

In this section we will define each task's the inputs, outputs, and their operation.

1) Shut Down/Emergency Brake or Stop

The task controls are steer, brake, throttle, gear, lift and tilt. The task is responsible to shut down the operations when the mission is completed and Emergency Brake or Stop if the strategy model detects that the wheel loader is close to hit the truck or any other object in the site or if the instability of the vehicle while travel is detected.

The inputs of this task are direction, velocity, acceleration, steer angle, gear, height and the angle of the bucket. This task gets signal either from other tasks directly if an emergent situation is detected by other tasks while executing them or state selector when the mission selected by the user is completed.

2) Back from the Truck/Dig Site

The task controls are steer, brake, and throttle. This task is responsible to back from the truck after dumping the load or to back from the dig site after filling the bucket. The wheel loader should back with enough distance and facing a direction which can enable the wheel loader to easily travel to the dig site after dumping the load in the truck or to the truck after filling the bucket in the dig site.

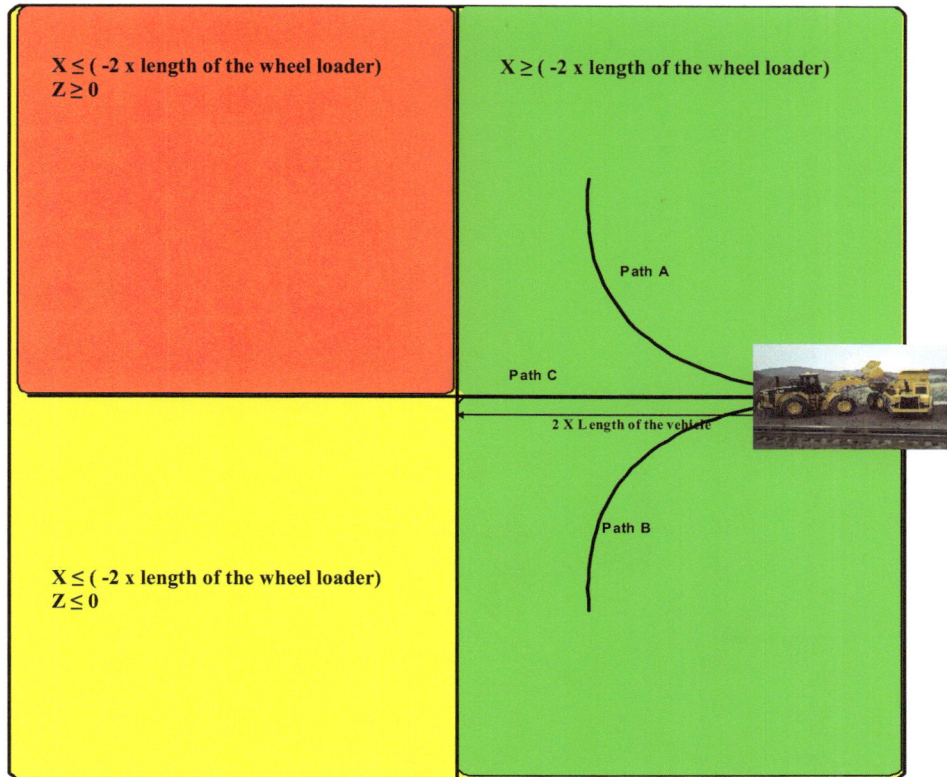

Fig. (10). Selecting the path criteria.

Depending on the position of start point (the truck after dumping the load or the dig site after filling the bucket) and the target point (the dig site or the truck) the task selects a path to take it while backing the wheel loader. The selection criteria as shown in Fig. (**10**). The task divides the field area into three parts starting from the start point where the X-axis direction is in the direction of the wheel loader and the Z-axis is the direction perpendicular to the wheel loader direction. X and Z are the coordinates of the target point with respect to the start point on X-axis and Z-axis. The first region, X is smaller than or equal twice the negative the length of the wheel loader. In this case the wheel loader backs choosing path C where the steer command equals to zero. The second region, X is greater than twice the negative the length of the wheel loader and Z is greater than zero. In this case the task choose path B for the wheel loader to back from truck/dig site. In the third and last region, X is greater than twice the negative of the length of the wheel loader but Z is smaller than zero.

The selection of path criteria was chosen according to the experimental testing and the operator behavior analysis of the experimental testing results. The velocity of the wheel loader while backing is chosen in a range with respect to operator comfort and to prevent spilling of the load if the bucket is full. The velocity is controlled using the throttle and brake commands. The accelerating and decelerating of the wheel loader has limits to prevent shaking of the vehicle, and spilling the load from the bucket. The acceleration and deceleration limits are chosen according to operator behavior analysis of the experimental testing. This task usually works in parallel with the gear selector that will be defined later. This task is selected by the state selector only in case the user selected a complete truck loading cycle or a part from it as a mission to be done.

The schematic of the task is shown in Fig. (**11**). The inputs to the task are truck/dig site position, direction, velocity and acceleration of the vehicle and outputs the steer command directly, and the desired velocity to throttle and brake controllers to control speed. The path selector chooses a path depending on truck/dig site position and direction of the wheel loader as explained. While the wheel loader following the selected path if it backed an enough displacement and the task is done the task gives signal to task # 3 to start. If the path is in a way to hit another object the task changes the direction to avoid hitting the object. If the tires are slipping and the controllers cannot prevent it due to ground conditions the task changes the direction to prevent slipping. If the wheel loader is very close to hit an object or detects an emergency the task gives signal to task #1 to start control.

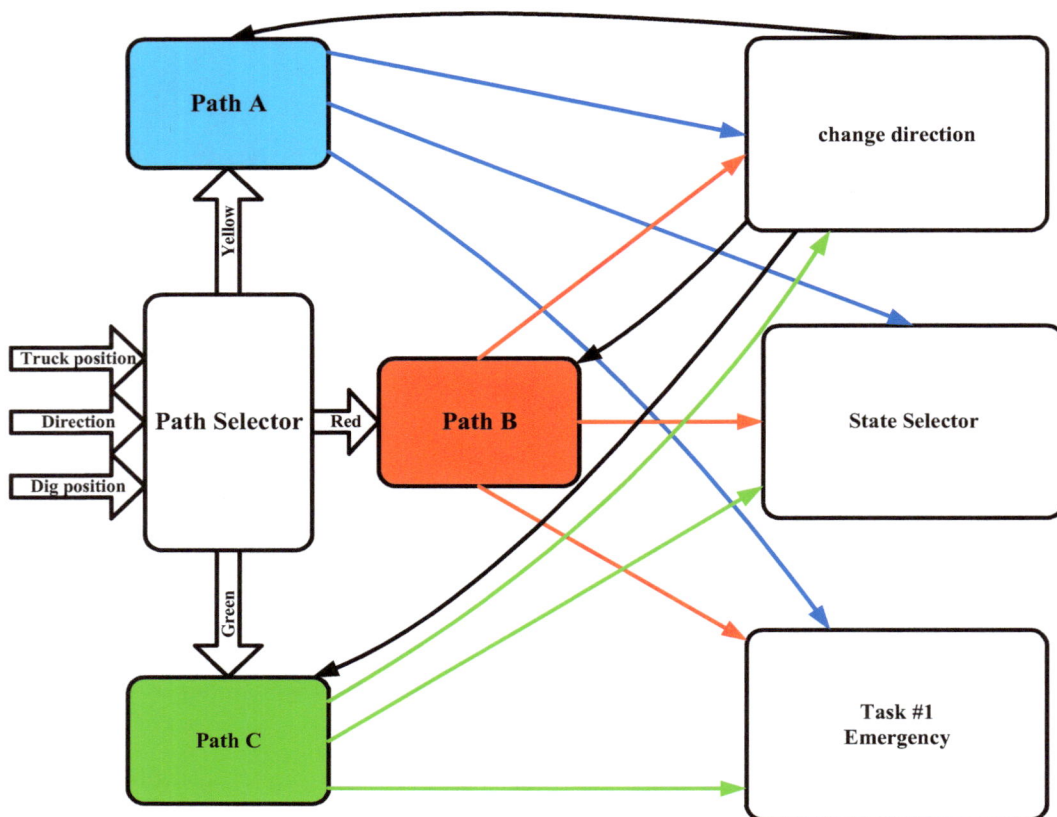

Fig. (11). Back from the truck/dig site schematic.

Go to Site/Truck

The task controls steer, throttle and brake controllers. The task is responsible to control the wheel loader to reach the target point (the truck to dump the load or the dig site to fill the bucket). The task controls the wheel loader to reach the truck/dig site perpendicular to its direction to ease dumping the load/filling the bucket and using the shortest distance, with a reasonable speed that allows the fast reach to the target point. The velocity of the wheel loader was chosen to prevent the spilling of the load if the bucket were full, and to prevent the instability of the vehicle. The acceleration and deceleration ranges were chosen to allow a comfortable control of the vehicle with respect to the operator. The velocity and the acceleration ranges were chosen depending on the operator behavior analysis while experimental testing. This task is selected if the user selected the complete truck loading cycle or a part of it and the evaluation test no 5.

The schematic of the task is shown in Fig. (12). The inputs are the target position, the current position, steer angle and the yaw velocity. This task works in parallel with gear selector task and lift/lower the bucket if needed. Using the input information the path controller selects the steer controller in parallel with the velocity control algorithm to control the wheel loader to reach a point 2 meters before the target point. After reaching this point the wheel loader goes straight till it reaches the target. This way helps the wheel loader to reach the target point in a direction perpendicular on the truck/dig site direction. While the vehicle is moving in straight direction the state selector gives signal to task # 4 (lift/lower the bucket) to start in parallel to ease the start of task # 5 (dump load/rack the bucket) or task # 6 (filling the bucket). After the target is reached, the state selector gives a signal to start the next task to be done.

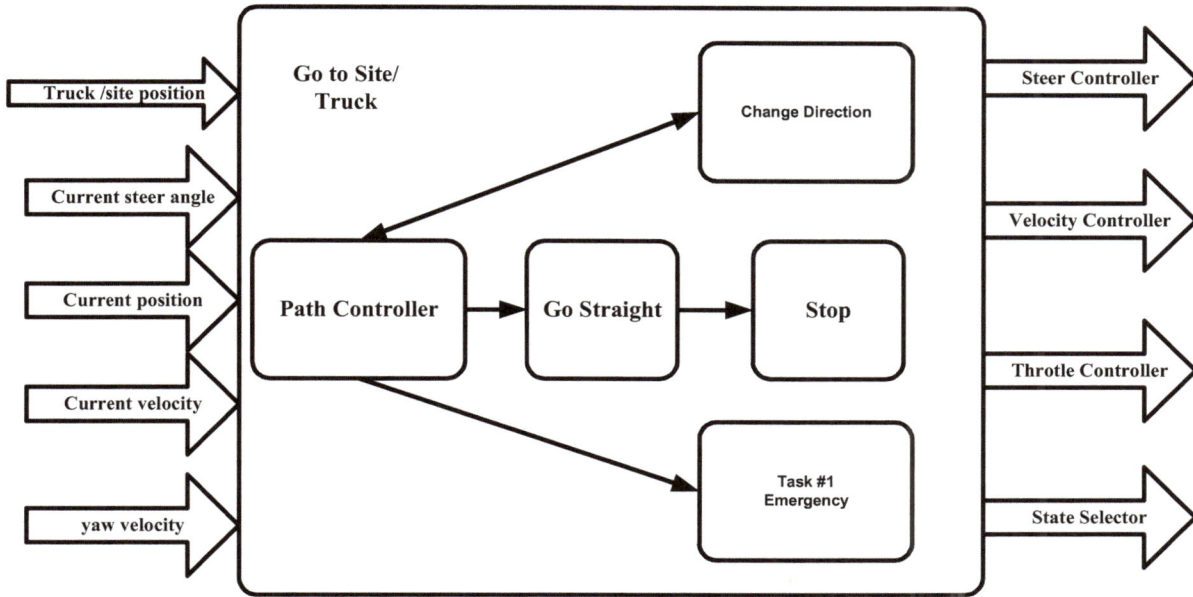

Fig. (12). Go to truck/dig site schematic.

The path controller continuously calculates the error between the current position and the target if it exceeds a certain limit which means the steer controller is not controlling the direction properly the task changes the direction by giving directly a steer command bypassing the steer controller to correct the direction then gives the control back to the steer controller. If the tires are slipping due to ground conditions the task change direction then goes back to the steer controller. When the task detects instability, very high speed, high acceleration or very close to hit an object the path controller gives signal to task #1 (emergency control) to take control.

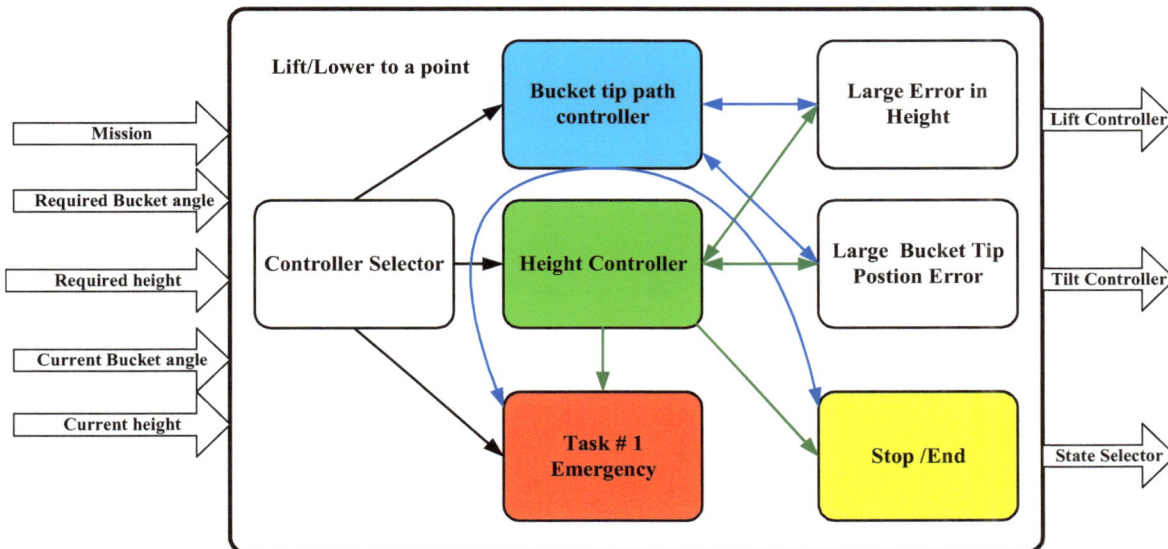

Fig. (13). Lift/lower the bucket schematic.

4) Lift/Lower the Bucket

This task controls the lift and tilt controllers which are the controllers for the hydraulic implement system in the model. It is responsible for the height and the angle of the bucket. This task is selected by the state selector in all missions that the operator model can do. The accuracy of this task determines the controllability of the wheel loader to a great extent.

The inputs are the mission, current bucket angle, current height, and required height and bucket angle in the case of evaluation tests. The controller selector selects either the bucket tip path controller or the height controller depending on the mission. The height controller is selected during the complete truck loading cycle. The bucket tip controller is selected in all evaluation tests.

After selecting the controller, the error between the current and the desired position and angle is continuously calculated. If the error exceeded a certain limit, this means the selected controller is not controlling the bucket properly. The task gives a direct lift/tilt command bypassing the controller to correct the direction then gives the control back to the selected controller.

If the bucket tip was close to hit another object while operation, the task gives signal to task#1 to take control. This task is selected to work alone in evaluation tests or parallel with the two previous tasks in the truck complete truck loading cycle.

5) *Dump Load/Rack Bucket*

This task also controls the lift and tilt motion. It is responsible to dump the load in the truck and rack the bucket after that or evaluation test number 3.

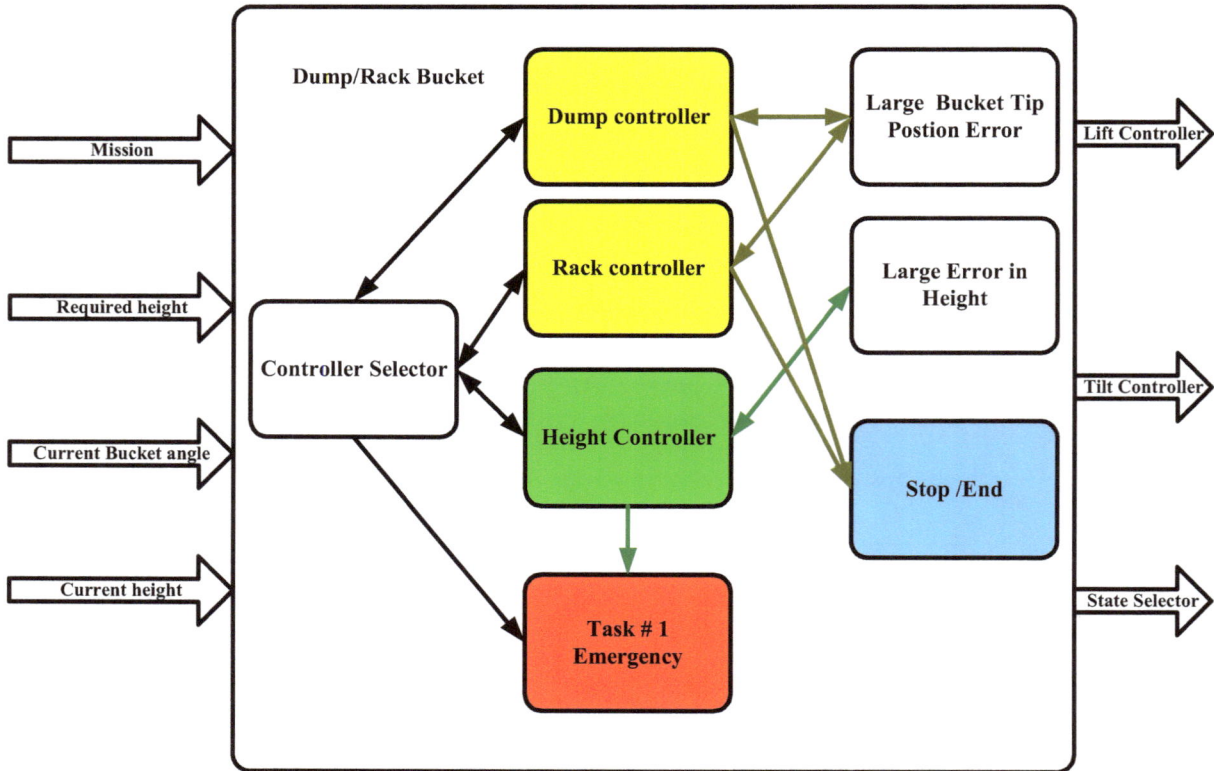

Fig. (14). Dump/rack the load schematic.

The schematic of this task is shown in Fig. (**14**). The inputs are mission, required height/truck height, current height and the current angle. The controller selector selects from height controller or dump controller or rack controller depending on the mission and the sequence of the tasks needed. The height controller is selected in parallel with either the dump controller or the rack controller.

The dump controller controls the bucket angle through tilt command to fully tilt out the bucket to dump the load completely without hitting the stops or producing unwanted high vibrations in the vehicle. The rack controller controls the bucket angle through tilt command to fully tilt in the bucket. The height controller works in parallel with both controllers to control the height of the bucket tip.

After selecting the controller, the error between the current and the desired position and angle is continuously calculated. If the error exceeded a certain limit, this means the selected controller is not controlling the bucket properly. The task gives a direct lift/tilt command bypassing the controller to correct the direction then gives the control back to the selected controller.

6) Gear Selector

This task controls the gear selection only. It works in parallel with all the tasks. The schematic for the task is shown in Fig. (15). The inputs are the direction, the current gear the current task, engine speed in revolutions per minute and the load.

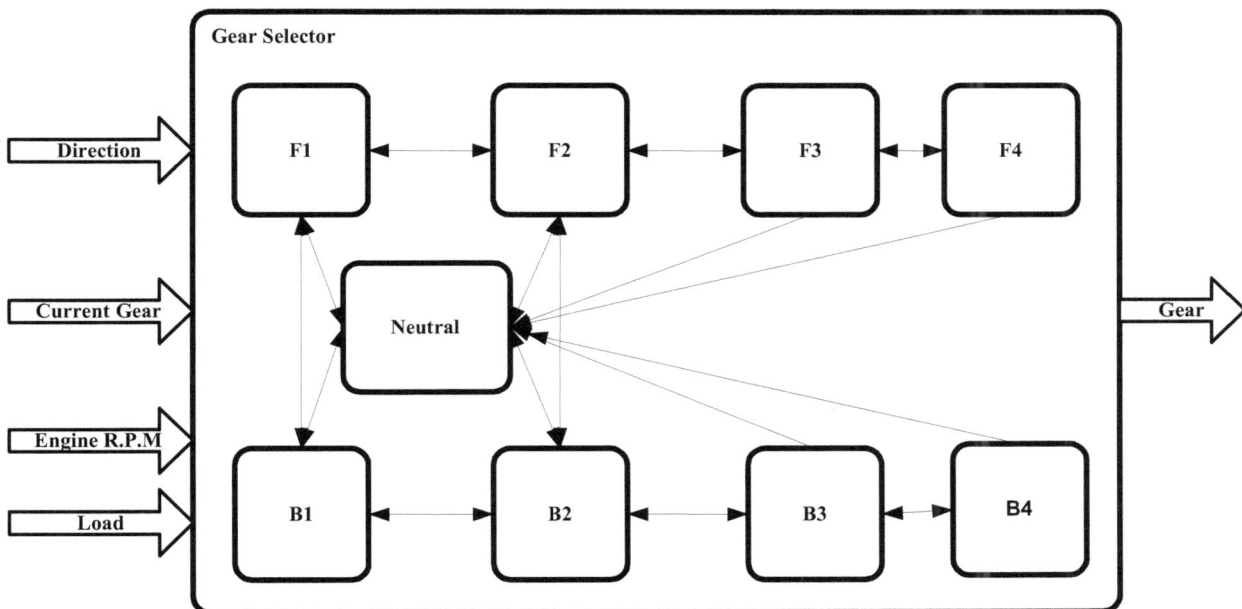

Fig. (15). Gear selector schematic.

For tasks number 3 and 4 the gear selector selects the neutral gear if they tasks were not done in parallel with other tasks. If other tasks are done in parallel the gear is selected according to the other task. For task number 1 which is the emergency stop the neutral gear is selected. The other tasks the gear is selected according to engine speed and load in the bucket.

While moving forward or backward the gear can start from neutral to first or second depending on the load in the bucket to prevent tire slipping. If the tire slips when high torque is applied a higher gear is selected. The gear selector can change the gear directly from any shift back to neutral.

7) Filling the Bucket

This task controls throttle, gear, steer, lift and tilt controllers. The task is responsible to fill the bucket with its payload. It is the main function of the wheel loader. The schematic of the task is shown in Fig. (16). The inputs are load, current steer angle, current bucket height and current bucket angle.

First the task checks for the direction of the vehicle. The wheel loader should be perpendicular on the target point to ease digging and filling the bucket. If it was not perpendicular the task steer right or left. Then the steer command is set to zero while filling the bucket.

The controller selector selects between a height controller and bucket angle controller to choose a reasonable penetration angle to optimize filling the bucket. The task selects the first gear and increases the throttle to increase the penetration force.

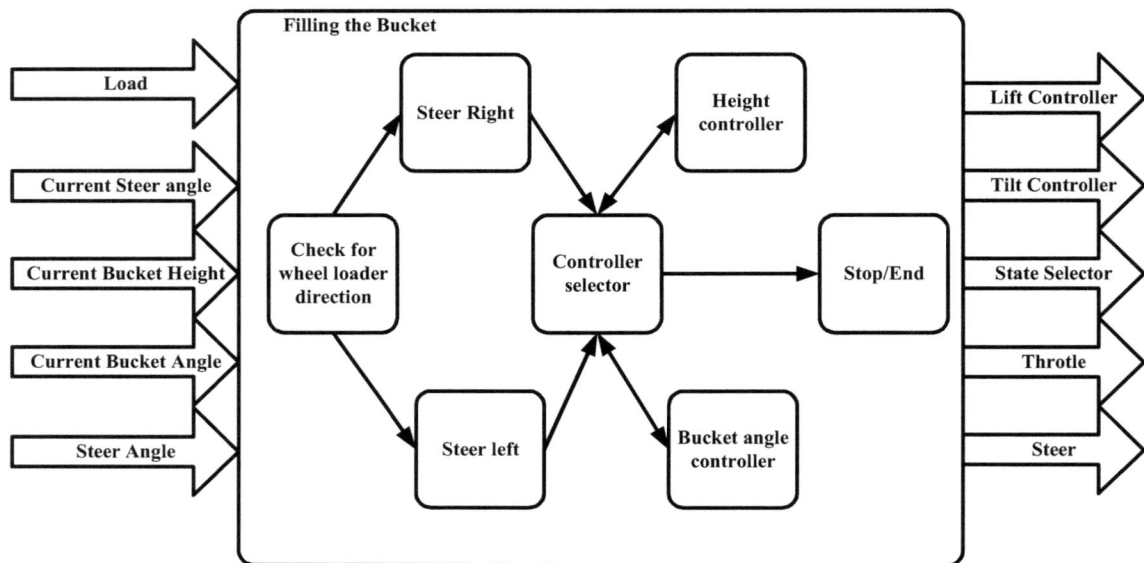

Fig. (16). Filling the bucket schematic.

The Controllers for Sub Tasks

In this section we define the controllers used to control steer, throttle, brake, lift and tilt to achieve the best performance of the wheel loader and allow the operator model to modify itself to control all models of the wheel loader.

The controllers used are based on neural networks; PID controllers were used in the first version (Elezaby *et al.*, 2008) but the problem with PID the tuning is that not enough information were known about the systems so they were tuned experimentally. First, the proportional term was increased from zero until the tilt function began to oscillate. It was then slightly reduced. Next an integral term was added to offset the steady state velocity errors that existed. An integral gain term quickly removed steady state errors without introducing excessive phase lag to the system. Finally, a derivative term was added in order to catch high frequency changes in the error signal. It worked with a good performance, but it worked only with one model and for every model tuning was required.

The neural network (MATLAB 2008); they can be trained offline to tune the controllers to a reasonable performance then online to modify itself to control various models of the wheel loader. Due to that fact that the command in the wheel loader models control the velocity output of the actuator, and the output of the strategy model is usually a desired position output therefore the error between the desired position output and the actual position output is fed to PID controller then the output is fed to the neural network controller. The neural network controller compares this input to the actual velocity of the actuator. The command is then determined depending on the error between this two values.

Training the neural network controllers takes three steps. The first step is done offline. A neural network model for the plant is trained to closely clone the plant performance. This procedure saves a lot of time in training the controller. An input is fed to the plant and the neural network plant model. And the error between the two responses is fed to a learning algorithm which modifies the weights and the biases of the neural network. In our controllers the neural network plant model was trained against the data recorded from experiments conducted on the wheel loader.

The second step is also done offline. A desired output is fed to the neural network controller and gives the control action to the neural network plant model. The response of the neural network plant model is compared with the response of the reference model. The error between the two responses is fed to a learning algorithm that modifies the weights and biases of the neural network controller till it gives an acceptable performance. The performance of the neural network controller is measured by the error between the plant response and the reference model.

The last step is the online training against the plant (shown in Fig. (**17**)). While operating the desired input is given to the neural network controller. The controller gives a control signal to the plant and the neural network plant model. The error between the plant and the neural network plant model responses are fed to a learning algorithm to modify the neural network plant model weights and biases. The error between the plant response and the reference model output is fed to a learning algorithm which modifies the neural network controller weights and biases if needed. This reference model controller is trained automatically to control new wheel loader models without any user modifications.

The neural networks used are feedforward backpropagation networks. Backpropagation was created by generalizing the Widrow-Hoff learning rule to multiple-layer networks and nonlinear differentiable transfer functions (Matlab Help, 2008). Input vectors obtained from actual machine tests and the corresponding target vectors are used to train a network until it can approximate a function, associate input vectors with specific output vectors, or classify input vectors in an appropriate way.

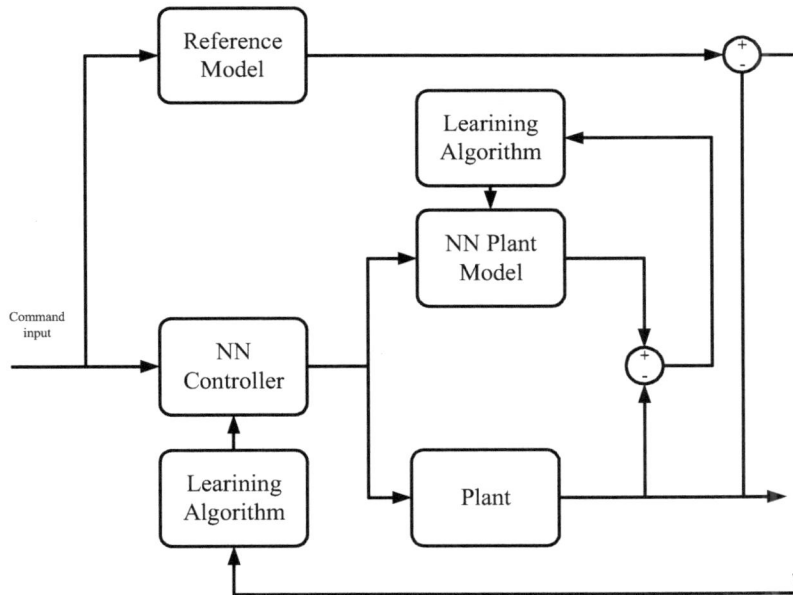

Fig. (17). Online training procedure *(from MATLAB 2008).*

THE MACHINE MODEL

The wheel loader model illustrated in Fig. (**18**) is a 3D body/3D tires model that simulates a medium size articulated type wheel loader. The wheel loader has an articulated type steering, which enables the vehicle to turn by articulating the front and rear frames about the hitch point. The wheel loader can be divided into four subsystems: 1) power train, 2) brakes, 3) steering, and 4) hydraulic implements. The power train consists of a power source which is typically a diesel engine. Power transmitted to a mechanical transmission *via* a torque converter which connects to differential drives and finally tires. Detailed dynamic model of the whole vehicle including its sub-systems, and tire-ground interaction and traction conditions, is a rather large engineering task (Fig. 20). Mathematical details of this part of the work is not discussed here.

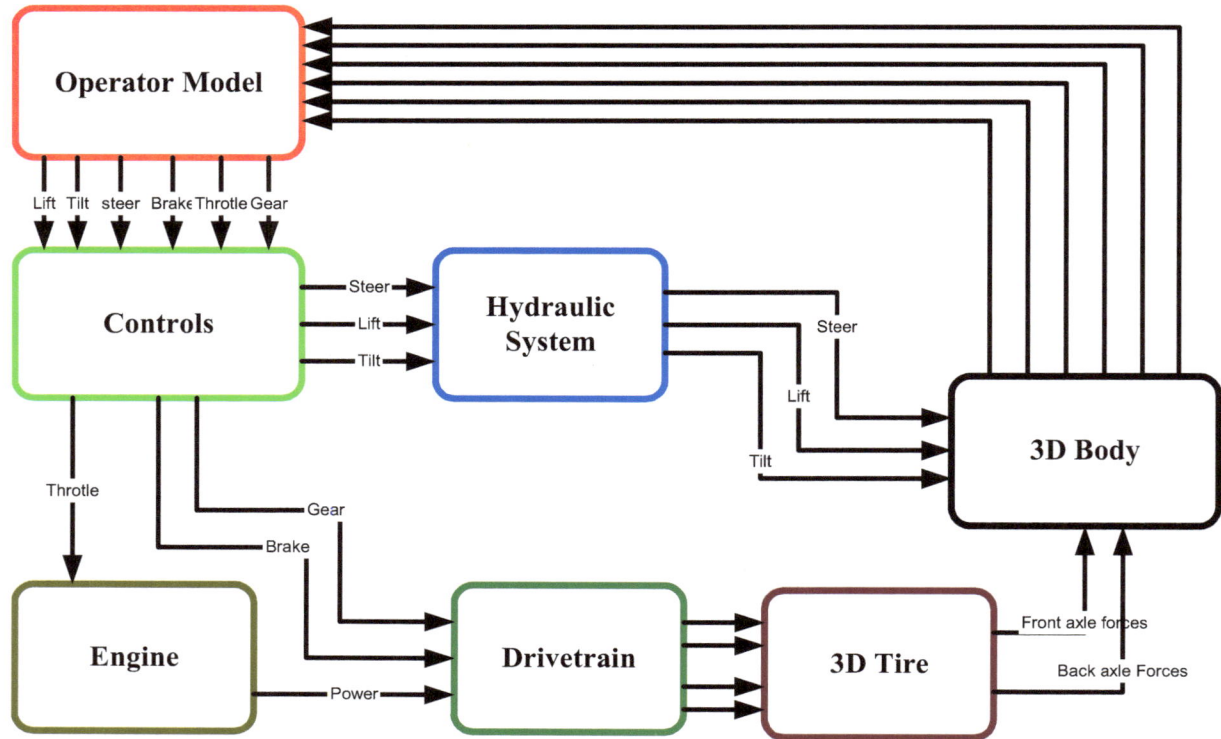

Fig. (18). The machine model schematic.

THE RESULTS

In this section we discuss the results of the operator model in simulation and comparing the results to the experimental results. First we will present the results of the complete truck loading cycle. The path chosen by the operator model for the wheel loader to complete a truck loading cycle is shown in Fig. (**19**). The wheel loader started in a position after dumped the load in the truck. According to the dig site position the operator model selected path B to back from the truck. After the wheel loader backed enough distance from the truck, the operator model selected the shown path to go to the dig site. While going to the site the operator model changed the bucket position to start digging and filling the bucket. When the bucket was filled with the load, the operator model selected path B to back from the site. Then the operator model took the path shown in the figure to go back to the truck. After assuring that the wheel loader is perpendicular to the truck, the operator model started giving the command to dump the load.

In Fig. (**20**) the human commands delivered from the operator model is shown. In the first task (Backing from the truck) the figure shows the application of a steer, throttle commands. After enough distance is reached the brake command is applied, also the gear command worked in parallel with the task. The first backward gear is engaged then neutral while braking. The second task selected is go to the site, the gear selector selected the first forward gear, the wheel loader started moving trying to keep a speed of 2 m/sec. The steer command was applied till a point 2 m behind the target point was reached, then the steer command stopped to allow the wheel loader to go straight to reach the target point perpendicular to the dig site. The wheel loader started digging using lift and tilt command. We can see the throttle was 100 % while digging to use maximum performance. After the bucket was filled, the wheel loader backed from the bucket. The human commands applied were steer, throttle and brakes. After backing enough distance, the wheel loader started going to the truck using steer throttle and brake command. The gear selector was working in parallel as shown in the figure. After reaching the truck, the operator model gave commands to dump the load in the truck.

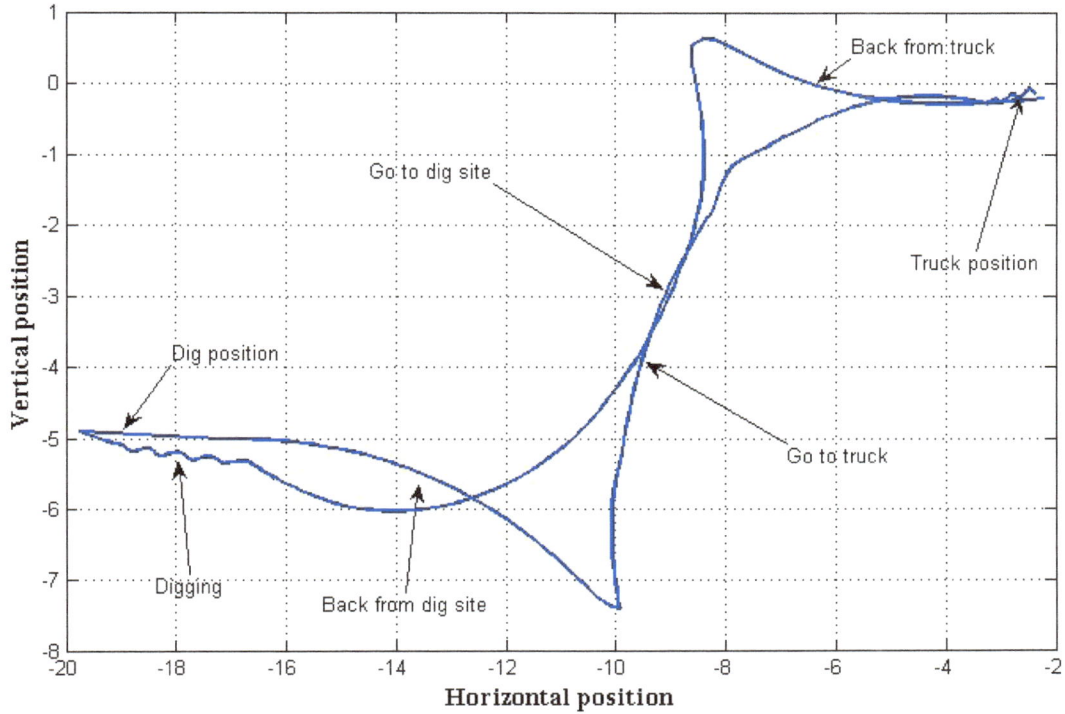

Fig. (19). Full cycle bucket tip path.

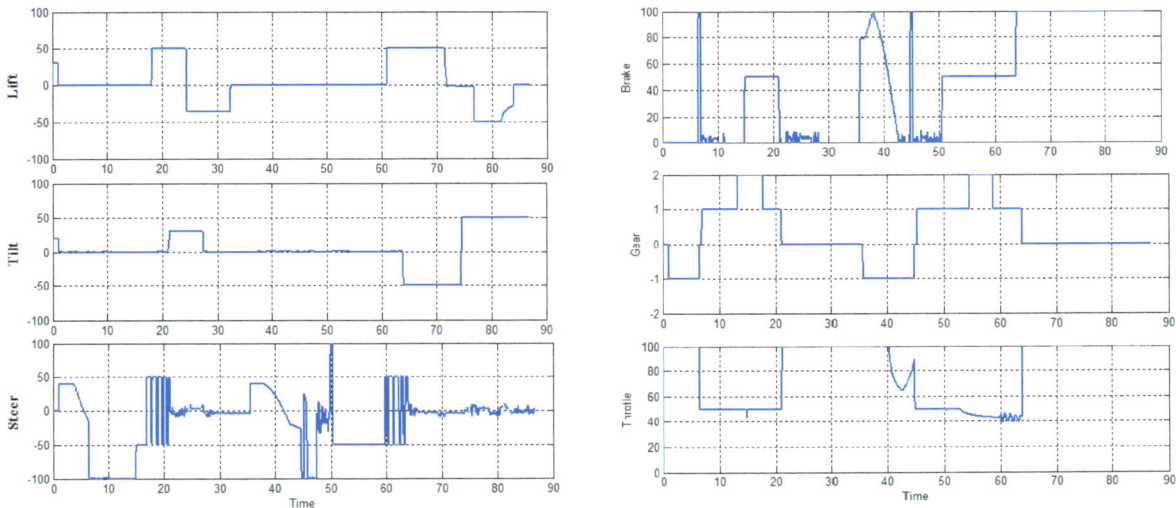

Fig. (20). Operator model commands.

In Fig. (**21**) shows a comparison between the experimental results and the simulation results of the operator model for the first evaluation test. The human operator started the test from up position then he performed three trails of the test. We can see that each trial is little different from one another. The operator model was apple to mimic the performance of the human operator with an error close in range in of the human error. The target was point at 3 meters high and 0.6 m easting. The human operator was able to reach a point 3.1357 m high and 0.6 m easting. The operator model reached a point of 3.1123m high and 0.6 m easting. The pattern of performing the test three times was achieved. The randomness in human performing was captured by the operator model as shown in the figure. The time consumed by the operator model was the approximately the same time taken by the human operator.

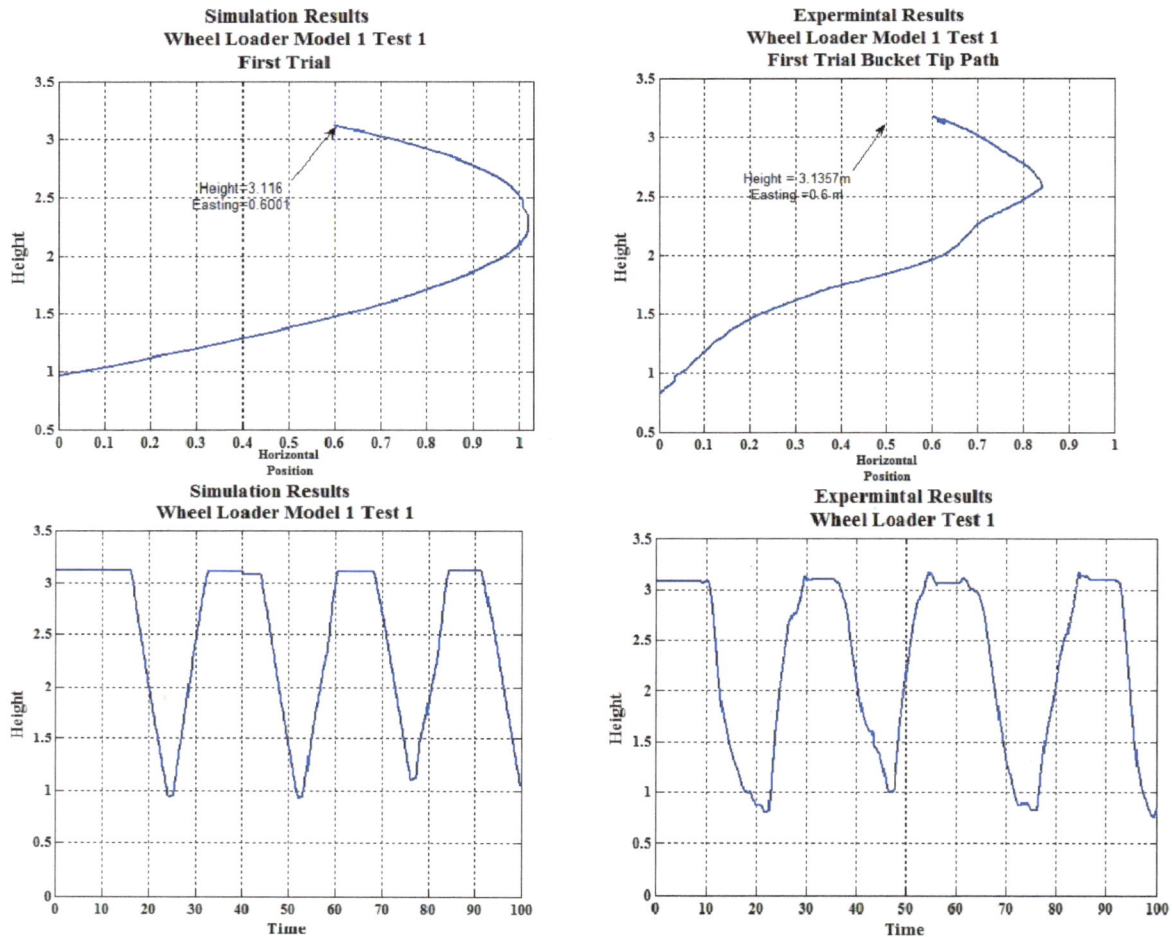

Fig. (21). Comparison between experimental and simulation results for test.

In Fig. (**22**) shows a comparison between the commands of the human operator and the operator model for the first evaluation test. We are comparing only the raising of the bucket part of the curve as this the goal of the test. In the lowering part the operator model was just giving a random command to lower the bucket. It shows a capture of the same pattern of the human operator commands. With the difference due to the fact as the operator model was controlling a virtual machine.

CONCLUSIONS

The virtual operator model is useful in evaluating the quality of a new machine design and predicting its performance in real applications, in a way very similar to the human operators would use and evaluate it. Using this tool, potential problems can be identified in early design stage, hence reducing the costly prototype testing stage of the machine development. In increasing use of computer modeling in mobile product development, there are three main components to creating a virtual dynamic reality of the whole system: 1. Machine model with all its sub-systems, 2. environment (loading conditions, pile of soil, and ground conditions etc), 3. operator model (the way the machine is actually used by real operators). This paper addressed the need on the last item.

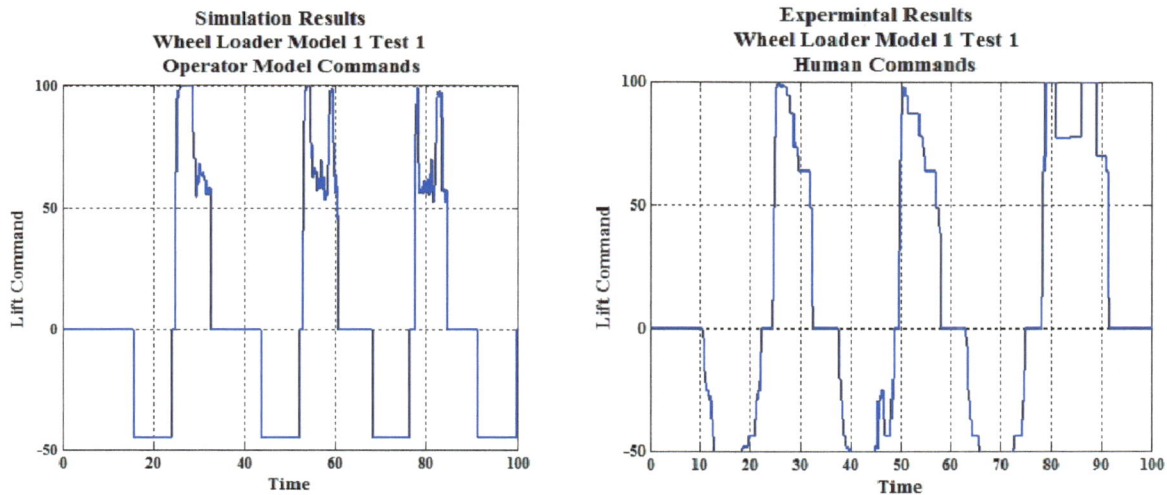

Fig. (22). Comparison between the commands of Test 1.

A virtual operator model with a human like performance was developed. The error range is less than 3 % with respect to the human performance in the truck loading cycle when compared the time of the cycle and efficiency. The virtual operator model was able to modify itself through online training to control various models and sizes of the wheel loader.

REFERENCES

Bengtsson, J. (2001). *Adaptive Cruise Control and Driver Modeling.* Licentiate Thesis, Lund Institute of Technology, Lund, Sweden.

Cetinkunt, S. (2006). *Mechatronics.* John Wiley & Sons, Inc.

Elezaby, A., Abdelaziz, M., & Cetinkunt, S. (2008). Operator Model for Construction Equipment. *2008 IEEE/ASME International Conference on Mechatronics and Embedded Systems and Applications*, October 12-15, Beijing, China.

Filla, R. (2005). An Event-driven Operator Model for Dynamic Simulation of Construction Machinery. *The Ninth Scandinavian International Conference on Fluid Power*, Linköping, Sweden, June 1-3.

Filla, R., Ericsson, A., & Palmberg J. (2005). Dynamic Simulation of Construction Machinery: Towards an Operator Model. *IFPE 2005 Technical Conference*, Las Vegas (NV), USA, pp. 429-438, March 16-18.

Filla, R. & Palmberg, J. (2003). Using Dynamic Simulation in the Development of Construction Machinery. *The Eighth Scandinavian International Conference on Fluid Power*, Tampere, Finland, May 7-9, *1*, 651-667.

Grant, P. (1994). Preparation of a Virtual Proving Ground for Construction Equipment Simulation. *IEEE Transactions on Industry Applications*, *30*(5), 1333-1338.

Lee, H. K., Barlovic, R., Schreckenberg, M., & Kim, D. (2004). Mechanical restriction versus human overreaction triggering congested traffic states. *Physical Review Letters*, 92(23), 238702-1-238702-4.

Macadam, C.C. (2003). Understanding and Modeling the Human Driver. *Vehicle System Dynamics*, 40(1-3), 101-134.

Marshall, J. A. (2001). Towards Autonomous Excavation of Fragmented Rock: Experiments, Modelling, Identification and Control. Master Thesis, Queen's University, Kingston, Ontario, Canada.

Mathworks, Inc. (2008). *Matlab Help.* Retrieved 2008, from http://www.mathworks.com

Shi, X. (1996). Experimental results of robotic excavation using fuzzy behavior control. *Control Engineering Practice*, 4(2), 145-152.

Singh, S. (1995). *Synthesis of Tactical Plans for Robotic Excavations.* Dissertation, Carnegie Mellon University, Pittsburgh, PA, USA.

Singh, S. (2002). State of the Art in Automation of Earthmoving, 2002. *Proceedings of the Workshop on Advanced Geomechatronics*, Sendai University, Japan.

Vogel, K. (2002). Modeling Driver Behavior – A Control Theory based Approach. Dissertation, Linköping University, Linköping, Sweden.

Index

A

Analytical redundancy 3, 11, 13, 15, 26-27

B

Brake By Wire 3-4, 7-8, 10

D

Dependability 40-44, 47, 49-51

Drive By Wire 3-5, 9-12, 14-15, 27

E

Electro-hydraulic 3-4, 7, 53, 57

Electromechanical 7-8, 34-35, 50

F

Fault Tolerant Control 3, 11, 13, 15, 26-27

Fault tolerant communication 3, 14

FDIA 11, 15, 21-22, 24-26

Fly by wire 12-14, 29

G

Generalized predictive 15, 19, 26

GPS 52-58, 66, 68, 75

H

Hardware-in-loop 20-21

N

Network 14, 27, 29-30, 34, 38, 42, 49, 70, 82-83

P

PID 67, 71, 82

S

Sliding Mode 3, 10, 15, 17-19, 26-27

Steer by wire 3, 9-10, 15, 22, 40-41, 49, 51, 52, 57-58

T

Throttle by wire 3, 6

V

Virtual 70, 74, 86-87

www.ingramcontent.com/pod-product-compliance
Lightning Source LLC
Chambersburg PA
CBHW041722210326
41598CB00007B/744